Biohybrid Catalysts for Alkene Metathesis

Von der Fakultät für Mathematik, Informatik und Naturwissenschaften der RWTH Aachen University zur Erlangung des akademischen Grades eines Doktors der Naturwissenschaften genehmigte Dissertation

vorgelegt von

Master of Science
Daniel Friedrich Sauer

aus

Düren, Deutschland

Berichter: Universitätsprofessor Dr. rer. nat. Jun Okuda

Universitätsprofessor Dr. rer. nat. Ulrich Schwaneberg

Tag der mündlichen Prüfung: 28.03.2017

AACHENER BEITRÄGE ZUR CHEMIE

(Aachener Beiträge zur Chemie ; Bd. 127)
Zugl.: Aachen, Techn. Hochsch., Diss., 2017

Daniel Friedrich Sauer
Biohybrid Catalysts for Alkene Metathesis

ISBN: 978-3-95886-187-9

Bibliografische Information der Deutschen Bibliothek
Die Deutsche Bibliothek verzeichnet diese Publikation in der Deutschen Nationalbibliografie; detaillierte bibliografische Daten sind im Internet über http://dnb.ddb.de abrufbar.

Das Werk einschließlich seiner Teile ist urheberrechtlich geschützt. Jede Verwendung ist ohne die Zustimmung des Herausgebers außerhalb der engen Grenzen des Urhebergesetzes unzulässig und strafbar. Das gilt insbesondere für Vervielfältigungen, Übersetzungen, Mikroverfilmungen und die Einspeicherung und Verarbeitung in elektronischen Systemen.

Vertrieb:

1. Auflage 2017
© Verlagsgruppe Mainz GmbH Aachen
Süsterfeldstr. 83, 52072 Aachen
Tel. 0241/87 34 34
www.Verlag-Mainz.de

Herstellung:

Druck und Verlagshaus Mainz GmbH Aachen
Süsterfeldstraße 83
52072 Aachen
www.DruckereiMainz.de

Satz: nach Druckvorlage des Autors
Umschlaggestaltung: Druckerei Mainz

printed in Germany
D 82 (Diss. RWTH Aachen University, 2017)

This work delineated here was carried out between October 2013 and December 2016 in the laboratories of Prof. Dr. Jun Okuda, at the Institute of Inorganic Chemistry of the RWTH Aachen University, Germany.

"Don't ever let someone tell you, you can't do something. Not even me. You got a dream, you got to protect it. People can't do something themselves, they want to tell you you can't do it. You want something, go get it. Period."

— **Pursuit of Happyness**

Table of Content

A. Introduction 1
    A.1. Concept of Artificial Metalloenzymes 1
    A.2. Artificial Metatheases 3
        A.2.1. Dative Anchoring Approach 4
        A.2.2. Supramolecular Anchoring Approach 5
        A.2.3. Covalent Anchoring Approach 7
    A.3. β-Barrel Proteins Nitrobindin and FhuA 12
    A.4. Aims and Scope of this Thesis 14
    A.5. References 15

B. Results and Discussion 18
    B.1. Metathease Based on the Transmembrane Protein FhuA 18
        B.1.1. Introduction 18
        B.1.2. Synthesis of Grubbs-Hoveyda Type Complexes with Varying Spacer Length 20
        B.1.3. Synthesis of the Biohybrid Conjugates Based on FhuA 21
        B.1.4. Analysis of the Biohybrid Conjugates based on FhuA 22
        B.1.5. Alkene Metathesis Reactions 27
        B.1.6. Summary and Conclusion 36
        B.1.7. Experimental Section 37
        B.1.8. References 46
    B.2. Metathease Based on the Soluble Protein Nitrobindin 48
        B.2.1. Introduction 48
        B.2.2. Synthesis and Characterization of Biohybrid Conjugates Based on NB 51
        B.2.3. Synthesis and Characterization of Biohybrid Conjugates Based on NB4exp 56
        B.2.4. Alkene Metathesis Reactions 62
        B.2.5. Summary and Conclusion 72
        B.2.6. Experimental Section 73
        B.2.7. References 78
    B.3. Whole-Cell Catalysis Based on β-Barrel Proteins FhuA and Nitrobindin 80
        B.3.1. Introduction 80
        B.3.2. Whole-Cell Metathesis Catalysts Based on the Transmembrane Protein FhuA 82
        B.3.3. Cell-Surface Display Technology to Generate a Whole-Cell System Based on Nitrobindin 85
        B.3.4. Summary and Conclusion 89
        B.3.5. Experimental Section 90

B.3.6. References 93
C. Summary 94
D. Appendix 97
D.1. List of Abbreviation 97
D.2. Curriculum Vitae 98
D.3. Publications 99
D.3.1. Publications in Peer-Reviewed Journals 99
D.3.2. Book Contributions 100
D.3.3. Patent Application 100
D.3.4. Conference Contributions 100
D.3.5. Other Publications 101
D.4. Acknowledgements – Danksagung 102

A. Introduction[a]

A.1. Concept of Artificial Metalloenzymes

Artificial metalloenzymes or biohybrid catalysts enable catalytic reactions or cascades in an aqueous environment or make existing catalysts more efficient. This interdisciplinary field between biotechnology and chemistry dates back to the late 1970's and attracted huge attention during the last decade.

In 1976, Kaiser and coworkers investigated the carboxypeptidase A (CPA).[1] This metalloenzyme contains a Zn^{2+} ion as active site in the native state and exhibits esterase and peptidase activity. By exchanging Zn^{2+} against Cu^{2+}, the natural activity disappeared completely, even though the structure of the enzyme and active site seems to be unchanged.[1] With incorporated Cu^{2+}, the artificial enzyme catalyzes the oxidation of ascorbate, like an oxidase. Two years later, Whitesides and coworkers incorporated an achiral, non-natural diphosphine rhodium catalyst into the protein avidin (Avi).[2] At the backbone of the catalyst, a biotin moiety was attached which undergoes supramolecular conjugation with the wild-type of Avi. This conjugation is nearly irreversible ($K_d = 10^{-12} - 10^{-15}$ M) and one of the strongest supramolecular interactions known in nature.[3] Upon conjugation, the generated catalyst achieved 44% *ee* in a hydrogenation reaction using H_2 as hydrogen source (Scheme 1).[2]

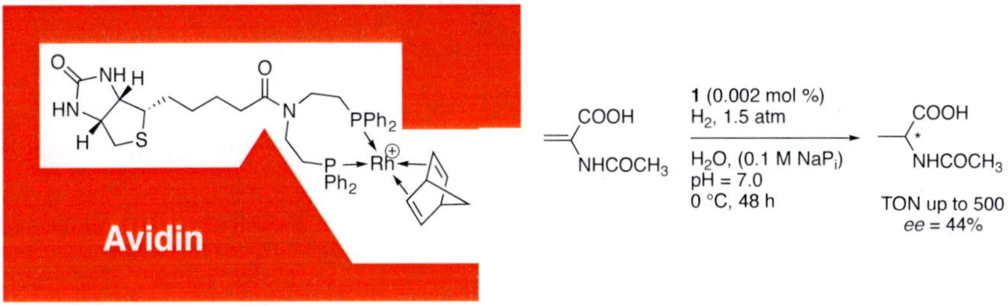

Scheme 1. Artificial hydrogenase by Whitesides and coworkers.[2]

It took over 25 years till the topic was picked up by Ward and coworkers.[4] The reason for the delay to study this system might be explained by a lack of methods and technologies for mutating, isolating and characterizing the proteins. The benefits of combining a protein with a (non-natural) metal is to alter the natural reactivity of an enzyme. As an example, Hayashi and

---

[a] Parts of this chapter have been published in: Sauer, D. F.; Gotzen, S.; Okuda, J. *Org. Biomol. Chem.* **2016**, *14*, 9174.

coworkers demonstrated that the oxygen transporting protein myoglobin catalyzes hydroxylation. The natural heme b cofactor was reconstituted with a synthetic manganese porphycene.[5-6] Furthermore, non-chiral catalysts show enantioselectivity inside the protein environment.[2,7-9] By incorporation of non-natural co-factors into a protein, the limited reaction scope in nature can be extended and new cascade reactions might become possible,[10-13] since many reactions catalyzed by those non-natural co-factors are biorthogonal. Several artificial metalloenzymes for a broad variety of reactions have been reported.[14-25]

A.2. Artificial Metatheases

Artificial metatheases are constructed by the incorporation of a Grubbs-Hoveyda type (GH-type) catalyst into a protein. The GH-type catalyst shows increased stability towards air and water[26-27] and offers huge potential for variation within the ligands of the first ligand sphere, mainly on the *N*-heterocyclic carbene (NHC) ligand (Figure 1).

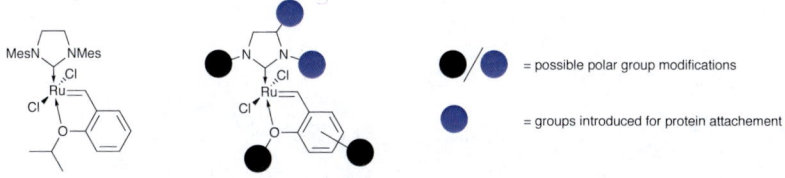

Figure 1. Modifications of GH-type catalyst. Black and blue spheres show the possible modification sites to achieve water-solubility.[27] Blue spheres show the sites where an anchoring unit was attached for protein immobilization.[25]

After Grubbs and coworkers published the first small molecule GH-type catalyst for metathesis in water,[28] Grela and coworkers investigated how the substitutions influence water-solubility and reactivity.[27,29-31]

GH-type catalysts with anchoring units can be connected to protein hosts by dative, supramolecular or covalent anchoring (Figure 2).

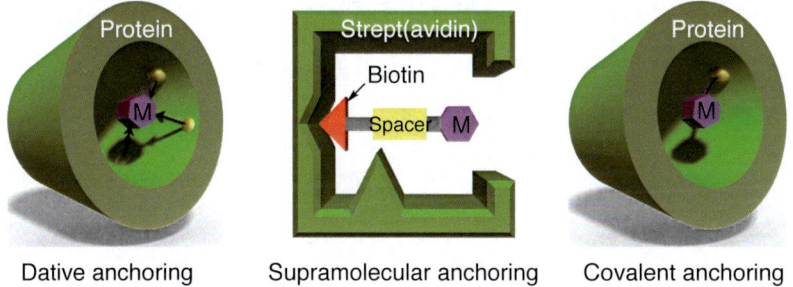

Figure 2. Illustration of the dative, supramolecular and covalent anchoring strategy.

All three anchoring strategies were used to construct artificial metatheases. Strategies to combine GH-type catalysts with proteins are presented below.

A.2.1. Dative Anchoring Approach

The dative anchoring strategy was demonstrated by Ward and coworkers who incorporated a Grubbs-Hoveyda type catalyst into the human carbonic anhydrase II (hCAII).[32] This molecular catalyst was equipped with an arylsulfonamide group that acts as inhibitor for the Zn^{2+} center in hCAII (Figure 3).

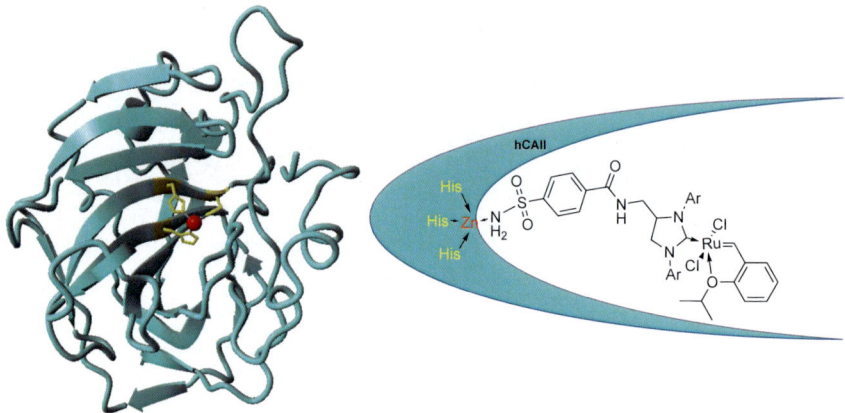

Figure 3. Crystal Structure of hCAII (left, Zn^{2+} in red; PDB code: 1ca2); Catalyst **2** (black) coordinated to hCAII (right).[32]

The binding affinity of catalyst **2** is relatively high (K_d = 205 nM).[32] This system was used for the ring-closing metathesis (RCM) reaction of diallylamine **3** (Table 1).

Table 1. Selected results for the RCM catalyzed by the metathease based on hCAII.[32]

		2 (1.0 mol %) 1.2 mol % of protein H_2O (0.1 M NaP_i) MCl_x, 37 °C, 4 h, $-C_2H_4$		
Entry[a]	hCAII	pH	(MCl_x) [M]	TON[b]
1	-	5	$MgCl_2$ (0.5)	85
2	-	6	$MgCl_2$ (0.1)	48
3	-	7	NaCl (0.154)	32
4	WT	7	-	14
5	WT	7	NaCl (0.154)	21
6	WT	7	NaCl (0.5)	32
7	L198H	7	-	22
8	L198H	7	NaCl (0.154)	28

[a] Conditions: c(**3**) = 1 mM, c(**2**) = 10 µM, c(hCAII) = 12 µM, V(DMSO) = 20 µL, total volume: 200 µL. [b] Determined by HPLC.

The strongest influences on the activity for this system are pH and salt concentration (Table 1, entries 5 and 6). This behavior is already well-known. Lower pH values improve the activity because other potentially coordinating ligands like hydroxyl groups or amino acid residues are protonated and therefore do not interfere in catalysis (Table 1, entries 1-3). The increased chloride concentration influences the stability of the ruthenium complex by suppressing water-chloride ligand exchange. This was shown by Matsuo et al. for water-soluble GH-type catalysts.[33] While the pH and the chloride concentration dominate the overall behavior of the system, the mutation introduced in the protein have only limited influence (Table 1, entries 7 and 8). Until now, this system is the only artificial metathease obtained by dative anchoring.

A.2.2. Supramolecular Anchoring Approach

Ward and coworkers used the established biotin-(strept)avidin system to construct an artificial metathease.[23] The GH-type catalysts were equipped with a biotin moiety to achieve supramolecular coordination (Figure 4). These biohybrid catalysts were investigated in the RCM reaction (Table 2).

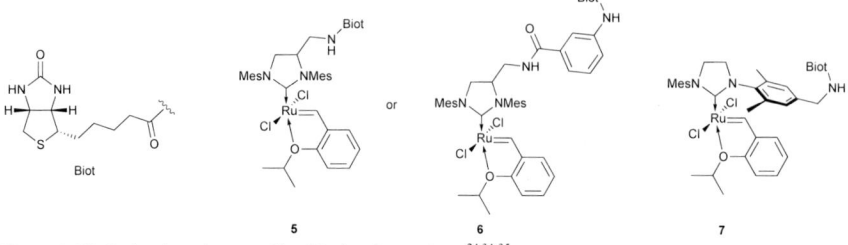

Figure 4. Biotinylated catalysts used by Ward and coworkers.[24,34-35]

Table 2. Optimized results for the RCM reaction catalyzed by an artificial metathease based on the biotin-(strept)avidin technology.[24,34-35]

Entry[a]	Complex	Substrate	Protein	Conv. (%)	TON
1	5	3	-	74	15
2	7	8	-	19	38
3	6	3	Sav-WT	71	14
4	5	3	Avi-WT	95	20
5	7	8	Sav-K121A	33	66

[a] c(**5** or **6**) = 0.73 mM, final volume 120 µL; c(**7**) = 0.05 mM, final volume 200 µL.

In first trials with streptavidin (Sav), the biohybrid conjugate showed low conversions of the water insoluble substrate **3** (TON < 1). After optimizing the reaction conditions, the biohybrid catalyst achieved TON comparable to the protein-free catalyst (Table 2, entries 1 and 3). Furthermore, the protein matrix was changed to avidin (Avi) and the spacer length was adjusted. This led to a slight increase in activity (Table 2, entry 4). However, the biggest increase in activity was achieved by using a more water-soluble substrate **8**. This led to an increase in activity for both, the protein-free and the biohybrid conjugated system, where the biohybrid system showed an increased activity compared to the protein-free system (Table 2, entries 2 and 5).

This system was transferred to a living-cell by expressing the host protein Sav into the periplasm of *E. coli* (Scheme 2).[24]

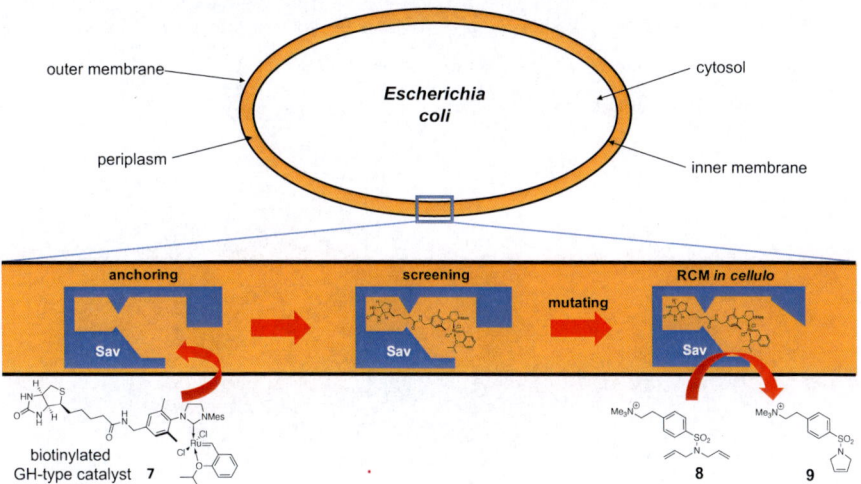

Scheme 2. Whole-cell approach demonstrated by Ward and coworkers.[24]

For comparison, whole-cells without Sav and with Sav expressed in the cytoplasm were produced. Catalyst **7** was introduced by incubation of the whole-cells in buffer solution containing the catalyst. After incubation, the unbound catalyst was removed by washing. The metal content was determined by ICP-AES revealing 26.000 Ru atoms per cell in the absence of Sav. With Sav expressed in the cytoplasm, 31.000 Ru atom per cell were detected. When the Sav is expressed into the periplasm, the Ru count increased to 80.000 atoms per cell. Using the Sav expressed in the cytoplasm, the RCM reaction of substrate **8** resulted in a maximum TON of approximately 500.000 per cell. This corresponds to a TON = 6 related to the metal. In the absence of Sav or with Sav expressed in the cytoplasm, conversion was not observed.[24] This shows in a proof of principle that the metathesis reaction is suitable for being performed in a whole-cell environment. Open questions are:

i) How is the catalyst transported into the cell?
ii) How are the ruthenium atoms bound to the cells that do not contain any Sav?
iii) How can substrate **8** reach the catalyst inside the cell, because its charged nature should prevent diffusion through the lipophilic membrane?
iv) What happens to the ethylene formed during the reaction?

A.2.3. Covalent Anchoring Approach

In 2012, Matsuo *et al.* reported the elegant construction of an artificial metathease by using supramolecular pre-coordination of the GH-type catalyst **10** within the cavity of the protein α-chymotrypsin.[36] This catalyst is equipped with a L-phenylalanyl moiety acting as inhibitor for this protease. Covalent bond formation is realized by nucleophilic attack of histidine H57, substituting the chloride in the backbone of the GH-type catalyst leading to the biohybrid conjugate **11** (Figure 5).

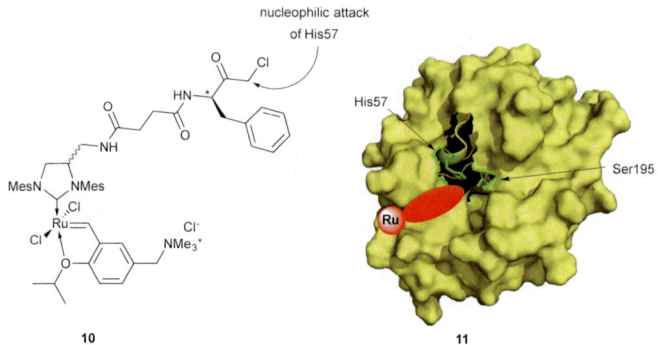

Figure 5. GH-type catalyst **10** and the artificial metathease **11** based on α-chymotrypsin (position of **10** is shown in red).[36]

The biohybrid conjugate **11** with covalently bound active center was used for the RCM reactions of three substrates and compared to the protein-free catalyst **10** (Table 3).

Table 3. Results of the RCM reactions catalyzed by **10** and **11**.[36]

Entry	Catalyst	Substrate	TON	Conv. (%)
1[a]	11	13	20	12.5
2[a,b]	10	13	14	8.8
3[c]	11	12	n.d.	n.d.
4[c,d]	10	12	< 2	< 5
5[b]	11	3	4	10
6[b,e]	10	3	12	30

[a] c(**13**) = 8.0 mM, c(catalyst) = 0.05 mM. [b] containing 10% (v/v) DMSO. [c] c(**12**) = 8.0 mM, c(catalyst) = 0.2 mM. [d] containing 10% (v/v) MeOH. [e] c(**3**) = 1.0 mM, c(catalyst) = 0.025 mM.

Compared to the protein free catalyst **10**, the biohybrid system **11** performs better with substrates that are soluble in water like the glycosylated substrate **13** (Table 3, entries 1 and 2). With substrates that do not dissolve in water like **3**, the water-soluble biohybrid catalyst is less active than the molecular catalyst without protein (Table 3, entries 5 and 6). The GH-type catalyst **10** showed only low conversion for the charged substrate **12**. The catalyst immobilized on a protein did not show conversion at all (Table 3, entries 3 and 4). This is explained by the positive charge within the polar cavity that leads to electrostatic repulsion of substrate **12**.[36] However, it was not shown by mutation of the cavity that the repulsion is the reason for the diminished activity of catalyst **11**.

Hilvert and coworkers covalently incorporated a GH-Type catalyst into the capsid structure of *M. jannashii small heat shock protein* 16.5 (MjHSP; Scheme 3).[37]

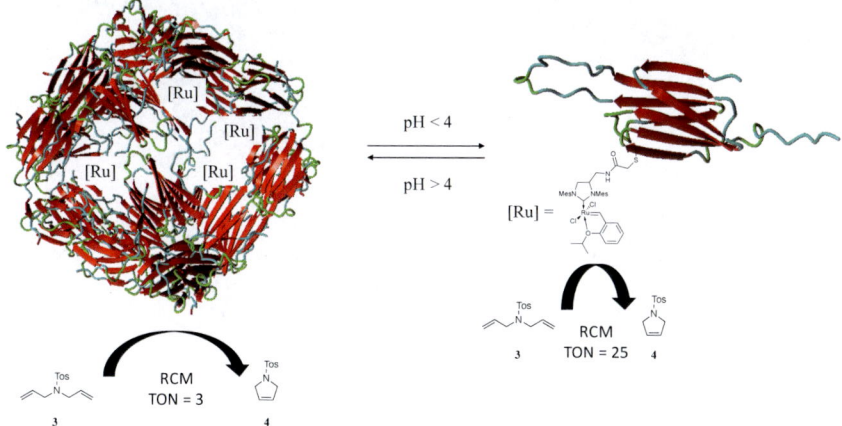

Scheme 3. Artificial metathease based on MjHSP (PDB code: 1shs).[37]

At pH > 4, the capsid structure of MjHSP is stable but the activity in the RCM of amine **3** is moderate (TON = 3). Adjusting the pH value to two breaks the capsid structure into smaller subunits that contain the catalyst and are more active in RCM (TON = 25).[37]

Klein Gebbink and coworkers utilized the lipase cutinase (**Cut**) as protein scaffold (Figure 6).[38]

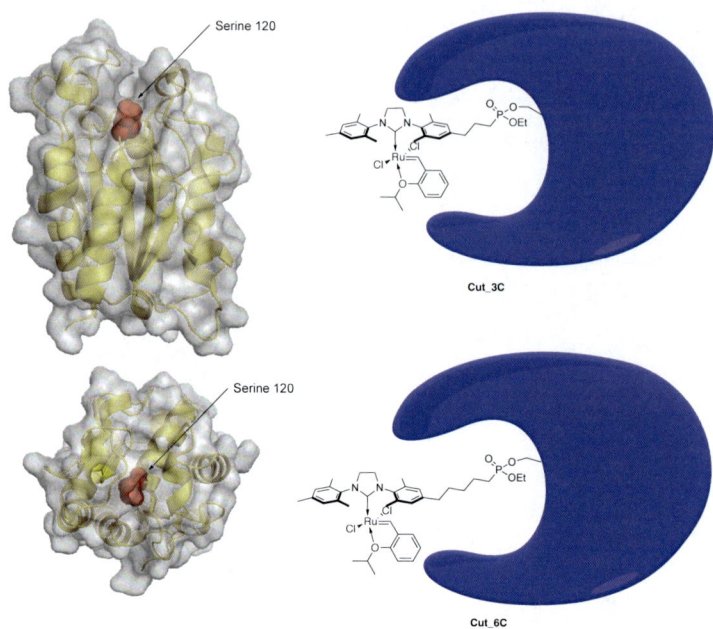

Figure 6. Artificial metathease based on the lipase cutinase (PDB code: 1cex).[38]

The GH-type catalysts were equipped with a phosphonate group (Figure 6). Phosphonate acts as inhibitor for the cutinase, forming a covalent bond between serine S120 and the catalyst. Two biohybrid catalysts **Cut_3C** and **Cut_6C** with different lengths of the linker were obtained (Figure 6).[38] These catalysts were probed in the RCM reaction of amine **3** and in the cross metathesis (CM) reaction of allyl benzene (Table 4).[38]

Table 4. RCM and CM catalyzed by the artificial metathease based on a lipase.[38]

Entry	Reaction	Catalyst	Yield [%]	TON	E/Z
1	RCM	**Cut_3C**	1	1	-
2	RCM	**Cut_6C**	84	17	-
3	CM	**Cut_3C**	2	1	63/37
4	CM	**Cut_6C**	> 99	20	36/64

RCM and CM reaction were carried out in a biphasic mixture using dichloromethane as organic phase. The linker length is a crucial for both reactions. A shorter linker only gave TON = 1 (Table 4, entries 1 and 3), a longer linker allowed much higher yields (Table 4, entries 2 and 4). The results are the same as those with the protein-free GH-type catalyst. This is the first example of a CM reaction with hybrid catalysts.[25]

A.3. β-Barrel Proteins Nitrobindin and FhuA

The proteins used to construct artificial metatheases contain charged residues within or at the edge of the corresponding cavity. These charged residues repel charged substrates[36] or result in unwanted coordination of the metal center to residues within the protein framework and therefore the reactivity diminishes. Recent investigations of an artificial metathease in a whole-cell environment[24,35] and in the reactivity based on the streptavidin technology revealed a strong influence of the position K121.[7-8,39] Mutation to a system with non-coordinating residues improved the activity significantly. Another example is the artificial metathease based on a variant of the β-barrel transmembrane protein *Ferric hydroxmate uptake protein component: A* (**FhuA**).[40-41] Immobilization of a GH-type catalyst catalyst lowered the activity in the ROMP reaction of a 7-oxanorbornene derivative by more than 50% compared to the protein-free system.[40] This was explained by the channel structure of the **FhuA**. In its native state, this protein is an iron transporter located in the outer membrane of *E. coli*.[42] To use this protein as scaffold for biohybrid catalysts, Schwaneberg and coworkers adapted the space within the channel structure by several mutations, especially by removing the cork domain that is responsible for the transport of iron.[43-44] An accessible cysteine (C545) was introduced to allow covalent anchoring of the metal catalysts.[40] So called TEV-cleavage sites were implemented to digest the protein into smaller fragments which can be analyzed by MALDI-TOF MS.[40-41] The final mutant investigated is named **FhuA_ΔCVF**[tev] or in short, **FhuA** (Figure 7). The hydrophilic character within the channel is maintained after removing the cork (Figure 8).

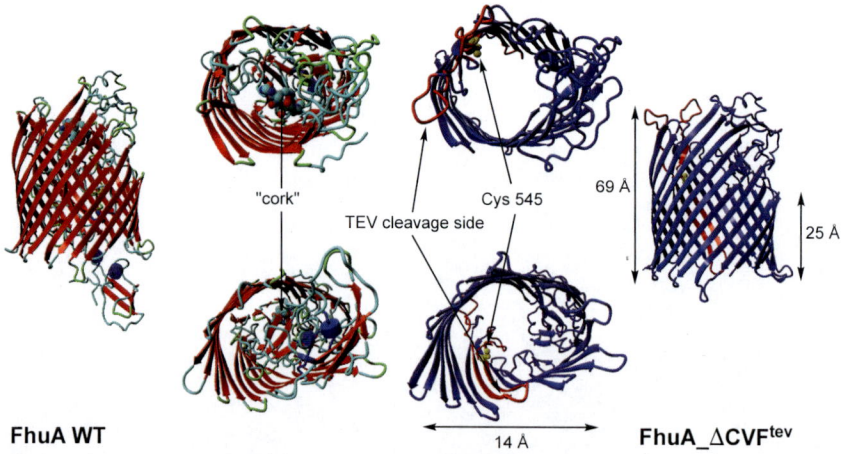

Figure 7. FhuA WT and FhuA_ΔCVF[tev].

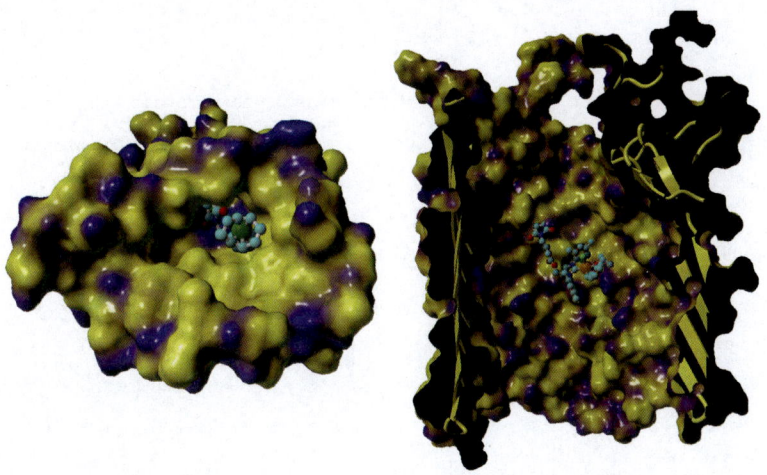

Figure 8. Left: **NB4_Rh** (Hayashi and coworkers[47]); Right: **FhuA_C3** (Okuda, Schwaneber and coworkers[40]). Hydrophobic residues are colored in yellow, hydrophilic residues are colored in blue; the atom shown in orange represents the Ru, the atom in green represents the Rh; hydrogens are omitted for clarity.

The soluble protein nitrobindin (**NB**) that naturally contains a heme unit provides a much smaller β-barrel structure than **FhuA** (22 β-sheets in **FhuA**; 10 β-sheets in **NB**).[45] Hayashi and coworkers obtained artificial metalloproteins by removing the heme and incorporating a rhodium cofactor.[46-47] **NB** provides a fully non-coordinative, hydrophobic environment (Figure 8). Hydrophobic residues are surrounding the rhodium center in **NB4_Rh**. The environment of the GH-type catalyst within **FhuA** is mixed with hydrophobic as well as hydrophilic residues. This makes **NB** promising for a GH-type catalyst and allows to investigate the difference between a hydrophobic and a hydrophilic cavity in β-barrel proteins.

A.4. Aims and Scope of this Thesis

This work aims at a better understanding of the behavior of the GH-type catalysts incorporated into a protein. The β-barrel proteins *Ferric hydroxamate protein component: A* (**FhuA**) and nitrobindin (**NB**) were chosen to anchor the GH-type catalyst. Based on earlier results for the transmembrane protein **FhuA**, the influence of the linker length is to be investigated by systematic shortening of the spacer attached at the GH-type catalyst. The protein nitrobindin shall be used because of the robust β-barrel structure. Non-coordinative environments are chosen to construct artificial metatheases. The results from metathesis reactions shall be compared to those obtained with other artificial metatheases.

Beside homogeneous systems with soluble or solubilized proteins, the behavior of the catalyst in a whole-cell environment will be explored, because these systems should be more stable and because simplified protein purification saves time and costs. The results obtained with artificial metatheases shall be transferred to other catalytic systems.

Within this collaborative research project, the synthesis of **FhuA** as solubilized protein or within whole-cells is carried out in the group of Prof. Ulrich Schwaneberg (RWTH Aachen University) by Dr. Marcus Arlt, M.Sc. Stephanie Mertens and M.Sc. Claudia Wichaltz. Soluble variants of the protein nitrobindin have already been designed and produced in the group of Prof. Takashi Hayashi (Osaka University) by Dr. Kazuki Fukumoto, Dr. Tomoki Himiyama, M.Sc. Kengo Tachikawa and B.Sc. Shunsuke Kato. The cell-surface display of nitrobindin and the expanded nitrobindin were designed and produced by Dr. Marco Bocola and M.Sc. Alexander Grimm (group of Prof. Ulrich Schwaneberg, RWTH Aachen University). The protein expression and production/purification is only mentioned briefly in the corresponding chapters and is the result of collaborative research.

A.5. References

(1) Yamamura, K.; Kaiser, E. T. *J. Chem. Soc., Chem. Commun.* **1976**, *0*, 830.

(2) Wilson, M. E.; Whitesides, G. M. *J. Am. Chem. Soc.* **1978**, *100*, 306.

(3) Green, N. M. In *Adv. Protein Chem.*; C.B. Anfinsen, J. T. E., Frederic, M. R., Eds.; Academic Press: **1975**; Vol. Volume 29, p 85.

(4) Collot, J.; Gradinaru, J.; Humbert, N.; Skander, M.; Zocchi, A.; Ward, T. R. *J. Am. Chem. Soc.* **2003**, *125*, 9030.

(5) Key, H. M.; Dydio, P.; Clark, D. S.; Hartwig, J. F. *Nature* **2016**, *534*, 534.

(6) Oohora, K.; Hayashi, T. In *Methods Enzymol.*; Vincent, L. P., Ed.; Academic Press: **2016**; Vol. Volume 580, p 439.

(7) Chatterjee, A.; Mallin, H.; Klehr, J.; Vallapurackal, J.; Finke, A. D.; Vera, L.; Marsh, M.; Ward, T. R. *Chemical Science* **2016**, *7*, 673.

(8) Zimbron, J. M.; Heinisch, T.; Schmid, M.; Hamels, D.; Nogueira, E. S.; Schirmer, T.; Ward, T. R. *J. Am. Chem. Soc.* **2013**, *135*, 5384.

(9) Podtetenieff, J.; Taglieber, A.; Bill, E.; Reijerse, E. J.; Reetz, M. T. *Angew. Chem. Int. Ed.* **2010**, *49*, 5151.

(10) Denard, C. A.; Huang, H.; Bartlett, M. J.; Lu, L.; Tan, Y.; Zhao, H.; Hartwig, J. F. *Angew. Chem. Int. Ed.* **2014**, *53*, 465.

(11) Denard, C. A.; Bartlett, M. J.; Wang, Y.; Lu, L.; Hartwig, J. F.; Zhao, H. *ACS Catal.* **2015**, *5*, 3817.

(12) Köhler, V.; Wilson, Y. M.; Dürrenberger, M.; Ghislieri, D.; Churakova, E.; Quinto, T.; Knörr, L.; Häussinger, D.; Hollmann, F.; Turner, N. J.; Ward, T. R. *Nat. Chem.* **2013**, *5*, 93.

(13) Okamoto, Y.; Köhler, V.; Ward, T. R. *J. Am. Chem. Soc.* **2016**, *138*, 5781.

(14) Lewis, J. C. *ACS Catal.* **2013**, *3*, 2954.

(15) Hoarau, M.; Hureau, C.; Gras, E.; Faller, P. *Coord. Chem. Rev.* **2016**, *308, Part 2*, 445.

(16) Hayashi, T.; Sano, Y.; Onoda, A. *Isr. J. Chem.* **2015**, *55*, 76.

(17) Heinisch, T.; Ward, T. R. *Eur. J. Inorg. Chem.* **2015**, *2015*, 3406.

(18) Yu, F.; Cangelosi, V. M.; Zastrow, M. L.; Tegoni, M.; Plegaria, J. S.; Tebo, A. G.; Mocny, C. S.; Ruckthong, L.; Qayyum, H.; Pecoraro, V. L. *Chem. Rev.* **2014**, *114*, 3495.

(19) Matsuo, T.; Hirota, S. *Bioorg. Med. Chem.* **2014**, *22*, 5638.

(20) Dürrenberger, M.; Ward, T. R. *Curr. Opin. Chem. Biol.* **2014**, *19*, 99.

(21) Bos, J.; Roelfes, G. *Curr. Opin. Chem. Biol.* **2014**, *19*, 135.

(22) Steinreiber, J.; Ward, T. R. *Coord. Chem. Rev.* **2008**, *252*, 751.

(23) Marchi-Delapierre, C.; Rondot, L.; Cavazza, C.; Ménage, S. *Isr. J. Chem.* **2015**, *55*, 61.

(24) Jeschek, M.; Reuter, R.; Heinisch, T.; Trindler, C.; Klehr, J.; Panke, S.; Ward, T. R. *Nature* **2016**, *537*, 661.

(25) Sauer, D. F.; Gotzen, S.; Okuda, J. *Org. Biomol. Chem.* **2016**, *14*, 9174.

(26) Garber, S. B.; Kingsbury, J. S.; Gray, B. L.; Hoveyda, A. H. *J. Am. Chem. Soc.* **2000**, *122*, 8168.

(27) Tomasek, J.; Schatz, J. *Green Chem.* **2013**, *15*, 2317.

(28) Jordan, J. P.; Grubbs, R. H. *Angew. Chem. Int. Ed.* **2007**, *46*, 5152.

(29) Lipshutz, B. H.; Ghorai, S. In *Olefin Metathesis*; John Wiley & Sons, Inc.: **2014**, p 515.

(30) Gawin, R.; Czarnecka, P.; Grela, K. *Tetrahedron* **2010**, *66*, 1051.

(31) Skowerski, K.; Szczepaniak, G.; Wierzbicka, C.; Gulajski, L.; Bieniek, M.; Grela, K. *Catal. Sci. Tech.* **2012**, *2*, 2424.

(32) Zhao, J.; Kajetanowicz, A.; Ward, T. R. *Org. Biomol. Chem.* **2015**, *13*, 5652.

(33) Matsuo, T.; Yoshida, T.; Fujii, A.; Kawahara, K.; Hirota, S. *Organometallics* **2013**, *32*, 5313.

(34) Lo, C.; Ringenberg, M. R.; Gnandt, D.; Wilson, Y.; Ward, T. R. *Chem. Commun.* **2011**, *47*, 12065.

(35) Mallin, H.; Hestericova, M.; Reuter, R.; Ward, T. R. *Nat. Protocols* **2016**, *11*, 835.

(36) Matsuo, T.; Imai, C.; Yoshida, T.; Saito, T.; Hayashi, T.; Hirota, S. *Chem. Commun.* **2012**, *48*, 1662.

(37) Mayer, C.; Gillingham, D. G.; Ward, T. R.; Hilvert, D. *Chem. Commun.* **2011**, *47*, 12068.

(38) Basauri-Molina, M.; Verhoeven, D. G. A.; van Schaik, A. J.; Kleijn, H.; Klein Gebbink, R. J. M. *Chem. Eur. J.* **2015**, *21*, 15676.

(39) Klein, G.; Humbert, N.; Gradinaru, J.; Ivanova, A.; Gilardoni, F.; Rusbandi, U. E.; Ward, T. R. *Angew. Chem. Int. Ed.* **2005**, *44*, 7764.

(40) Philippart, F.; Arlt, M.; Gotzen, S.; Tenne, S.-J.; Bocola, M.; Chen, H.-H.; Zhu, L.; Schwaneberg, U.; Okuda, J. *Chem. Eur. J.* **2013**, *19*, 13865.

(41) Sauer, D. F.; Bocola, M.; Broglia, C.; Arlt, M.; Zhu, L.-L.; Brocker, M.; Schwaneberg, U.; Okuda, J. *Chem. Asian J.* **2015**, *10*, 177.

(42) Ferguson, A. D.; Hofmann, E.; Coulton, J. W.; Diederichs, K.; Welte, W. *Science* **1998**, *282*, 2215.

(43) Güven, A.; Fioroni, M.; Hauer, B.; Schwaneberg, U. *J. Nanobiotechnol.* **2010**, *8*, 14.

(44) Tenne, S.-J.; Schwaneberg, U. *Int. J. Mol. Sci.* **2012**, *13*, 2459.

(45) Bianchetti, C. M.; Blouin, G. C.; Bitto, E.; Olson, J. S.; Phillips, G. N. *Proteins: Struct., Funct., Bioinf.* **2010**, *78*, 917.

(46) Onoda, A.; Fukumoto, K.; Arlt, M.; Bocola, M.; Schwaneberg, U.; Hayashi, T. *Chem. Commun.* **2012**, *48*, 9756.

(47) Fukumoto, K.; Onoda, A.; Mizohata, E.; Bocola, M.; Inoue, T.; Schwaneberg, U.; Hayashi, T. *ChemCatChem* **2014**, *6*, 1229.

B. Results and Discussion

B.1. Metathease Based on the Transmembrane Protein FhuA[a]

B.1.1. Introduction

The reported artificial metatheases were constructed utilizing soluble proteins (*cf.* chapter A). The first artificial metathease based on an extracted membrane protein was introduced in 2013.[1] Biohybrid catalysts based on the **FhuA_ΔCVF**[tev] (**FhuA**) and GH-type catalysts were used in the ROMP reaction of the 7-oxanorbornene derivative **16** (Scheme 4).

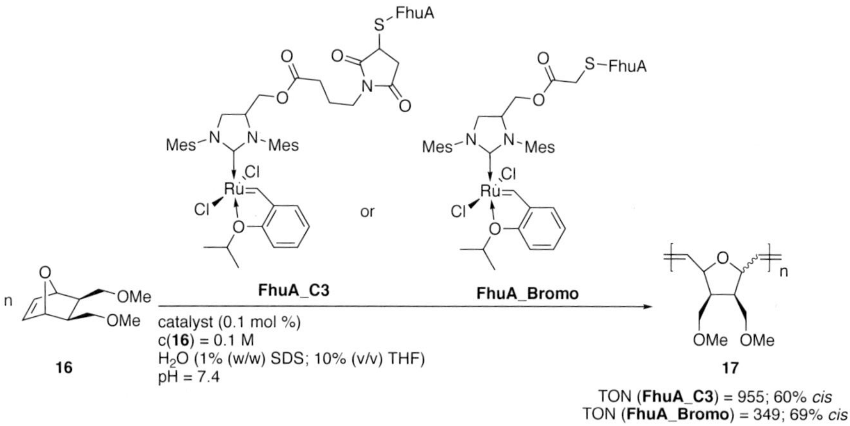

Scheme 4. ROMP of **16** with artificial metatheases based on **FhuA**.[1]

GH-type catalysts with a maleimide or a bromoacetyl anchoring unit were coupled to **FhuA**, resulting in the biohybrid conjugates **FhuA_C3** and **FhuA_Bromo**. Despite similar coupling efficiencies (> 90%), different activities and selectivity were observed (Scheme 4).[1] Conjugates based on **FhuA** require a detergent to be solubilized. These conjugates were used together with the detergent sodium dodecyl sulfate (SDS) that disturbs the natural β-barrel structure, what creates the so-called partially folded structure.[1-2] **FhuA_C3** is more active (TON = 955) than **FhuA_Bromo** (TON = 349). The *cis*-selectivity within the polymer is 60% for

---

[a] Parts of this chapter have been published in: Sauer, D. F.; Bocola, M.; Broglia, C.; Arlt, M.; Zhu, L.-L.; Brocker, M.; Schwaneberg, U.; Okuda, J. *Chem. Asian J.* **2015**, *10*, 177. Zhu, L.; Arlt, M.; Liu, H.; Bocola, M.; Sauer, D. F.; Gotzen, S.; Okuda, J.; Schwaneberg, U. In *Bio-Synthetic Hybrid Materials and Bionanoparticles: A Biological Chemical Approach Towards Material Science*; The Royal Society of Chemistry: **2015**, p 57.

FhuA_C3; **FhuA_Bromo** with the shorter spacer leads to a higher *cis* content (69%). Renaturing the β-barrel structure with the refolding reagent poly(ethylene)-*block*-poly(ethylene glycol) (PE-PEG, 0.125 mM) decreased the catalytic activity of the **FhuA_C3** to TON = 365 and made **FhuA_Bromo** completely inactive.[1] The RCM reaction of the benchmark substrate *N*-tosyl diallylamine **3** was attempted with **FhuA_C3**, but conversion was not observed due to protein precipitation. In the RCM reaction of the water-soluble 2,2-diallylpropane-1,3-diol, a TON = 35 was observed.[3]

B.1.2. Synthesis of Grubbs-Hoveyda Type Complexes with Varying Spacer Length

The maleimide thiol conjugation proved to be a promising anchoring approach.[1,4] To vary the spacer length, different linker units **L1** and **L2** were synthesized according to the procedure described by Christmann and coworkers (Scheme 5).[5]

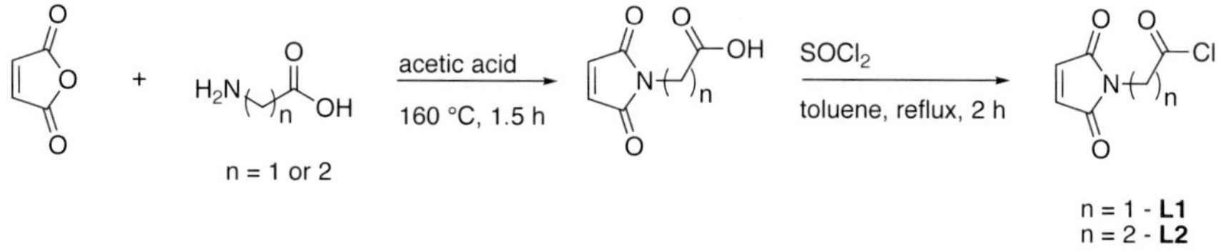

Scheme 5. Linker synthesis according to Christmann and co-wokers.[5]

The linking units were subsequently attached to the GH-type catalyst **COH** (Scheme 6).

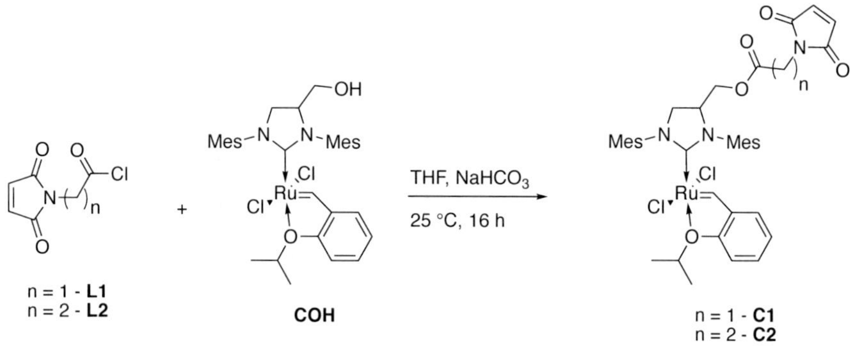

Scheme 6. Synthesis of GH-type catalysts **C1** and **C2**.

B.1.3. Synthesis of the Biohybrid Conjugates Based on FhuA

The new synthesized catalysts **C1** and **C2** as well as the previously reported **C3**[1] catalyst were attached to the **FhuA_ΔCVF^{tev}** (FhuA) variant (Scheme 7).

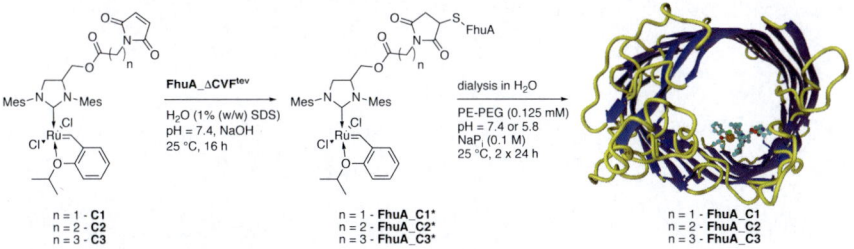

Scheme 7. Synthesis of the biohybrid conjugates based on **FhuA**.

Each catalyst **C1-3** was attached to the protein in sodium dodecyl sulfate (SDS) solution which made the position C545 more accessible. The β-barrel structure was renatured by dialysis. The block copolymer PE-PEG with a molecular weight of M_n = 2250 g/mol was used. This polymer consists of polyethylene (PE) and polyethylene glycol (PEG). The hydrophobic PE units interact with the hydrophobic part of the β-strands stabilizing the structure like the natural membrane does. The PEG units make the polymer water-soluble and keep the whole protein-detergent assembly in solution. PE-PEG and SDS form micelles in solution.[6-7] 2-Methylpentane-2,4-diol (MPD) was additionally used as refolding reagent. This small molecular reagent refolds the **FhuA** by interacting with the hydrophobic membrane area.

B.1.4. Analysis of the Biohybrid Conjugates based on FhuA

The artificial metatheases based on **FhuA** were analyzed by ThioGlo fluorescence titration of the thiol at C545. Mass spectrometry (MS) techniques were used and CD spectroscopy to confirm structural integrity.

In fluorescence titration, ThioGlo is attached to the protein scaffold by maleimide thiol coupling like the GH-type catalyst. This leads to fluorescence of the ThioGlo-protein adduct. This analysis was performed in the group of Prof. Schwaneberg by M.Sc. Stephanie Mertens, M.Sc. Claudia Wichlatz or by Dr. Marcus Arlt. Figure 9 shows the results of the fluorescence measurements.

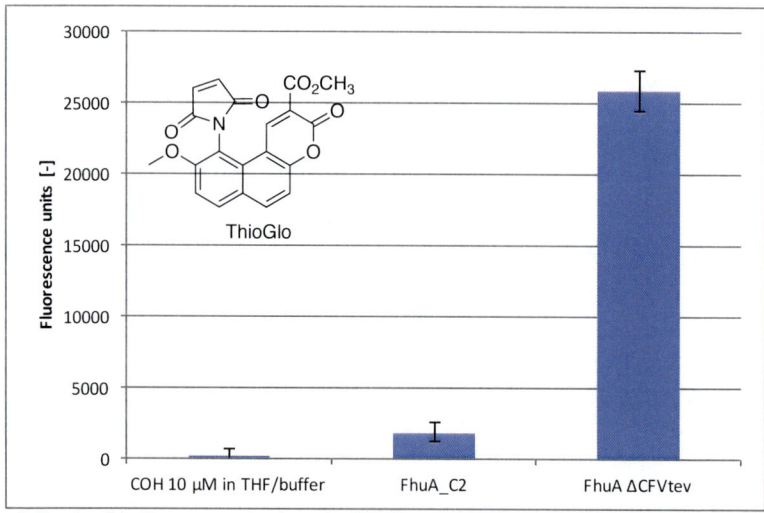

Figure 9. Results of the fluorescence titration using ThioGlo.

The **COH** catalyst used as background showed no fluorescence (Figure 9, left bar). As expected, pure **FhuA** shows a high fluorescence (Figure 9, right column). If a GH-type catalyst was attached to the protein before reaction with ThioGlo, the fluorescence was very weak (Figure 9, middle column). It shows that the thiol groups required for anchoring of ThioGlo are occupied. The ratio of the last two signals gives the coupling efficiency for each catalyst. The values for **C1-3** are higher than 90% and show successful anchoring.

Structural integrity was confirmed by CD spectroscopy (Figure 10).

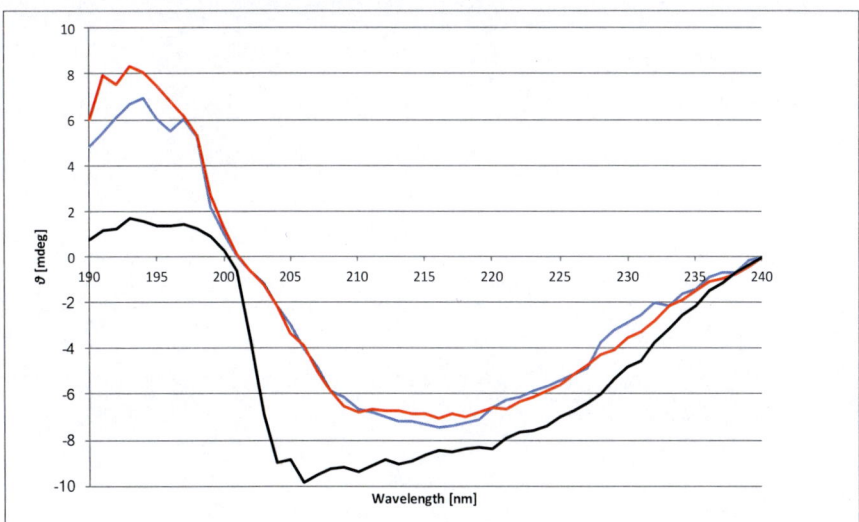

Figure 10. CD spectra for **FhuA_ΔCVF^{tev}** refolded in PE-PEG (red) and MPD (blue) and partially folded in SDS (black).

Figure 10 shows CD spectra for **FhuA** refolded in PE-PEG (Figure 10, red), MPD (Figure 10, blue) and partially folded in SDS (Figure 10, black). MPD and PE-PEG give very similar CD spectra showing the typical features of a β-barrel with a maximum around 195 nm and a minimum at 215 nm.[1-2,8] The corresponding spectrum in the presence of SDS shows significant different shape. Predicting the structure precisely is not possible.

MALDI-TOF MS is a suitable method to prove that a catalyst is covalently attached to a protein, but the high molecular weight of **FhuA** (63.5 kDa) cannot be determined precisely. This protein first had to be digested by a TEV protease into three smaller fragments, where the smallest fragment (ca. 5.9 kDa) contains the cysteine C545 with the attached catalyst.[1,8-9] **FhuA_Bromo** was analyzed previously by MALDI-TOF MS,[1] but the expected mass of the protein fragment with the metal co-factor could not be observed. The maleimide based catalysts **FhuA_C1-C3** are investigated by the same method as **FhuA_Bromo**. The refolded biohybrid conjugates were subjected to digestion in the research group of Prof. Schwaneberg by Dr. Marcus Arlt. The protein fragments were analyzed by MALDI-TOF MS (Figure 11).

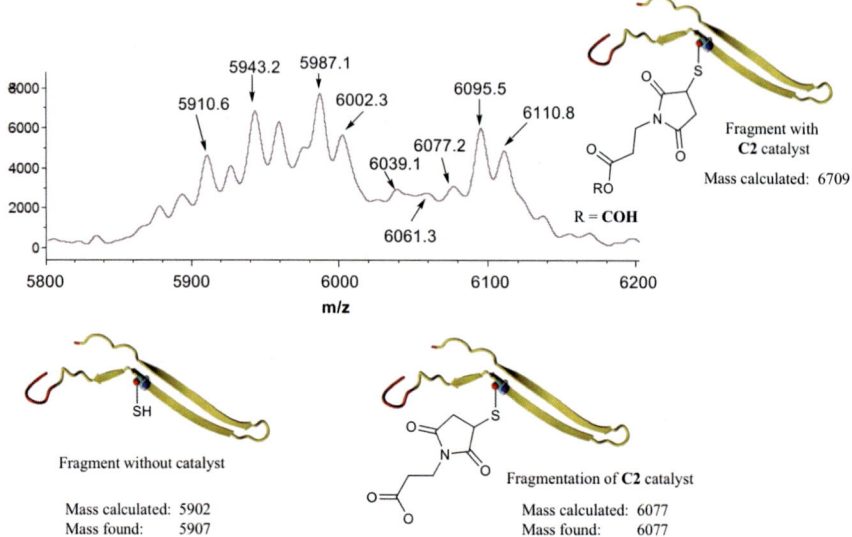

Figure 11. MALDI-TOF MS analysis of **FhuA_C2**.

The calculated mass of the fragment that contains **C2** catalyst (6709 Da) was not observed, but the spectrum clearly indicates that **C2** is covalently attached to the cysteine (Figure 11). The extremely weak signal for the apoprotein of 5902 Da corresponds to the results obtained by fluorescence titration with ThioGlo. The peak at m/z = 6077 is assigned to the product of ester cleavage at the backbone of the catalyst. The peak at m/z = 6095 is due to added water at the maleimide moiety. The biohybrid conjugates **FhuA_C1** and **FhuA_C3** give very similar results.[9]

To investigate if the ester is fragmented during digestion, during coupling or during ionization MS techniques were used. The catalysts **C1-C3** were transformed into model compounds by reaction with cysteine **Cys** (Scheme 8).

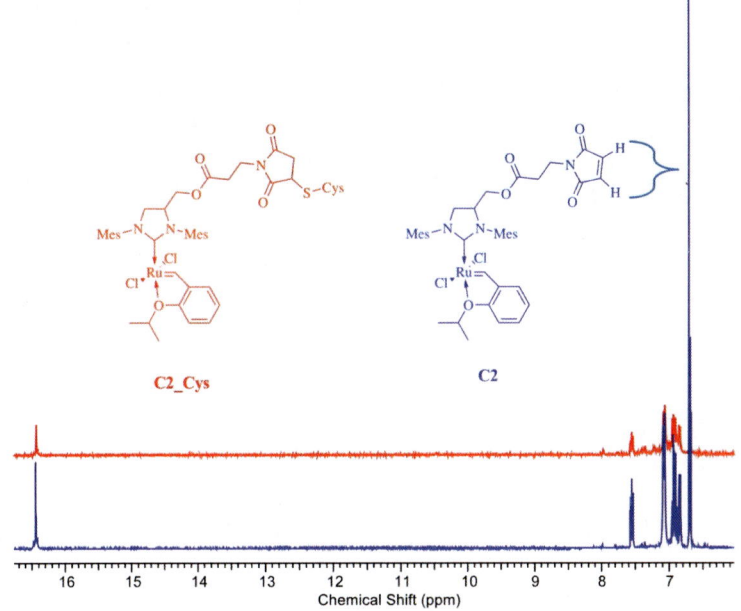

Scheme 8. Addition of cysteine **Cys** to the GH-type catalysts **C1-C3**.

The conjugation of the cysteine works smoothly in water under basic (pH = 8) conditions. The reaction took longer under neutral conditions and was not observed under acidic conditions. In THF, catalytic amounts of a base like *N*-methyl morpholine (NMM) were needed. The reaction was monitored by ^{1}H NMR spectroscopy (Figure 12).

Figure 12. ^{1}H NMR spectra (400 MHz, in CD$_2$Cl$_2$, 23 °C) of **C2** and **C2_Cys**. The signal at δ 6.6 ppm is assigned to the maleimide double bond. The aliphatic regions of the spectra are omitted for clarity.

The ^{1}H NMR spectrum of the isolated complex **C2_Cys** is similar to that ^{1}H NMR spectrum of catalyst **C2**. The missing signal at δ 6.6 ppm shows that covalent anchoring occured. Signals in the aliphatic region of **C2_Cys** show the formation of diastereomers upon conjugation to **Cys** (Scheme 8). If NaP$_i$ buffer had been used during the conjugation of **C1-C3** to **FhuA** in water, the catalyst precipitated. The increased salt concentration in the aqueous phase may have led to lower solubility of the catalyst.

^{1}H NMR spectroscopy confirmed that all cysteine conjugated catalysts retain their linking unit after conjugation. The ester units within **C1-C3** were not cleaved during coupling. MS analysis of catalysts **C1-C3** and **C1-C3_Cys** did not show the expected molecular mass. The dominant signal for all catalysts was the fragment after cleavage of the NHC-Ru bond. Fragmentation of the ester was also observed. Only with a lower ionization energy the corresponding masses of the catalysts could be observed.[8] However, the relative intensity was below 5% (for ESI spectrum and fragmentation, *cf.* B.1.7. Scheme 12). Similar findings were obtained with the corresponding cysteine conjugates. Fragments that correspond to products of McLafferty rearrangement at the *N*-boc group were observed with the typical isotopic pattern of the [RuCl$_2$] fragment (for ESI spectrum and fragmentation, *cf.* B.1.7. Scheme 13). These results agree with the earlier results on biohybrid conjugates.

B.1.5. Alkene Metathesis Reactions

B.1.5.1. ROMP Reaction with FhuA Based Biohybrid Catalysts

ROMP reaction of oxanorbornene **16** was performed with the **FhuA** biohybrid catalysts and the refolding reagents PE-PEG and MPD. Table 5 shows results together with those obtained with the protein free catalyst and with the cysteine conjugated catalysts.

Table 5. ROMP reaction of **16** in the presence of different refolding reagents.

Entry[a]	Catalyst	Refolding reagent	Conv.[b] [%]	cis/trans[b]	TON[c]
1	COH	PE-PEG	99	50/50	990
2	FhuA_C1	PE-PEG	41	58/42	555
3	FhuA_C2	PE-PEG	24	56/44	325
4[1]	FhuA_C3	PE-PEG	37	56/44	365
5	C1_Cys	PE-PEG	96	48/50	960
6	C2_Cys	PE-PEG	94	47/53	940
7	C3_Cys	PE-PEG	46	48/52	460
8	COH	MPD	99	50/50	990
9	FhuA_C1	MPD	45	57/43	610
10	FhuA_C2	MPD	25	56/44	340
11	FhuA_C3	MPD	40	58/42	540
12	C1_Cys	MPD	90	48/50	900
13	C2_Cys	MPD	94	50/50	940
14	C3_Cys	MPD	51	48/52	510

[a] $c(16) = 0.1$ M, THF 10% (v/v) [b] Determined by ^{1}H NMR spectroscopy in CDCl$_3$. [c] Based on the results using ThioGlo titration and BCA assay.

The protein free catalyst **COH** achieved full conversion within 68 h reaction time with a catalyst loading of 0.1 mol %. The *cis/trans* selectivity was 50/50 (Table 5, entry 1). The refolded protein catalysts showed approximately 2-3 times lower conversions, where the shorter linker of **FhuA_C1** showed higher conversion than the longer linker of **FhuA_C3**. The *cis* products were slightly favored (Table 5, entries 2-4). Like the protein free catalyst **COH**, the cysteine conjugates showed full conversion with *cis/trans* selectivities of ca. 50/50 (Table 5, entries 5-7). The cysteine did not influence the selectivity (Table 5, entries 12-14; for runs in THF, *cf.* B.1.7. Table 9; for runs in aqueous SDS solution Table 10 and for runs in aqueous THF solution Table 11). As expected, the refolding reagent MPD gave the same results, because it should not influence the active site within the β-barrel structure (*cf.* Table 5, entries 1-7 with entries 8-14).

B.1.5.2. RCM Reaction with FhuA Based Biohybrid Catalysts

FhuA_C1 was probed in RCM of *N*-tosyl diallylamine **3** (Table 6).

Table 6. RCM of **3** using catalyst FhuA_C1.

$$\underset{3}{\text{Tos-N(allyl)}_2} \xrightarrow[\substack{pH = 5.8 \\ \text{THF 10\% (v/v), Detergent} \\ 40\,°C,\ 750\ \text{mbar},\ 12\ h}]{\text{catalyst},\ H_2O\ (0.1\ M\ NaP_i)} \underset{4}{\text{Tos-N-pyrroline}}$$

Entry	Catalyst	Detergent[a]	Conv. [%]	TON
1[b]	FhuA_C1	PE-PEG	-[c]	-
2	FhuA_C1	MPD	0[c]	-
3	FhuA_C1*	SDS	1[c]	10
4[c,d]	FhuA_C3	PE-PEG	-	-
5[e]	GH@Avidin	-	> 99	20
6[f]	GH@MjHSP	-	> 99	25
7[g]	GH@α-chymotrypsin	-	12	4
8[h]	GH@Lipase	-	84	17
9[i]	GH@hCAII	-	32	32

[a] c(PE-PEG) = 0.125 mM; SDS 1% (w/w); c(MPD) = 50 mM. [b] Protein precipitation was observed. [c] Determined by ^{1}H NMR spectroscopy in CDCl$_3$. [d] RCM was performed with substrate **3** (50 mM) in an oxygen-free aqueous sodium phosphate buffer solution (0.1 M) / dry THF 10% (v/v) containing PE-PEG (0.125 mM) at 25 °C for 68 h.[3] [e] RCM was performed with substrate **3** (15.21 mM) in acetate buffer (0.1 M) containing MgCl$_2$ (0.5 M) and DMSO 17% (v/v) at 40 °C for 16 h.[10] [f] RCM was performed with **3** (5 mM) in HCl buffer (10 mM) containing ^{t}BuOH 8% (v/v) at 45 °C for 12 h.[11] [g] RCM was performed with **3** (1 mM) in KCl buffer (0.1 M) containing DMSO 10% (v/v) at 37 °C for 2 h.[12] [h] RCM was performed with **3** (1 mM) in acetate buffer (0.1 M) containing DCM 5% (v/v) at 25 °C for 20 h.[13] [i] RCM was performed with **3** (1 mM) in phosphate buffer (0.1 M) containing NaCl (0.5 M) and DMSO 10% (v/v) at 37 °C for 4 h.[14]

The refolded biohybrid conjugate **FhuA_C1** did not catalyze the RCM reaction of *N*-tosyl diallylamine **3**. In case of PE-PEG as refolding reagent, protein precipitation was observed. In case of MPD as refolding reagent, conversion was not observed (Table 6, entries 1-2). **FhuA_C1*** in SDS converted only 1% of substrate **3** (TON = 10). This has no driving force to enter the hydrophilic channel structure due to its hydrophobic properties, but reaches the metal site more easily after the β-barrel structure of the protein is disrupted using SDS. Other metatheases gave better conversions (Table 6, entries 5-9). It is noted that the reaction conditions for the **FhuA** based metathease differ for the other reported metatheases.

The water-soluble substrate **18** gave higher conversion (Table 7).

Table 7. RCM of diol **18**.

Entry	Catalyst (loading mol %)	Detergent[a]	Conv.[b] [%]	TON
1	FhuA_C1 (0.1)	PE-PEG	3	30
2	FhuA_C1 (1.0)	PE-PEG	44	44
3	FhuA_C1 (0.1)	SDS	4	40
4	FhuA_C1 (1.0)	SDS	80	80
5	FhuA_C1 (0.1)	MPD	3	30
6	FhuA_C1 (1.0)	MPD	40	40
7[3]	FhuA_C3 (0.1)	PE-PEG	3	35

[a] c(PE-PEG) = 0.125 mM; SDS 1% (w/w); c(MPD) = 50 mM. [b] Determined by ^{1}H NMR spectroscopy in CDCl$_3$.

FhuA_C1 showed 80% conversion with a catalyst loading of 1 mol % in SDS solution (Table 7, entry 4). Using PE-PEG or MPD as refolding reagent, moderate conversion of 44% or 40% was observed, respectively. (Table 7, entries 2 and 6). A reduced catalyst loading of 0.1 mol % decreased the conversion to 3-4%, independent of the surfactant (Table 7, entries 1, 3, 5). The results are like those obtained with catalyst **FhuA_C3** (Table 7, entry 7).[3]

B.1.5.3. CM Reaction with FhuA Based Biohybrid Catalysts

Cross metathesis (CM) is used in organic chemistry to transform terminal double bonds into internal double bonds by eliminating ethylene. The reverse reaction with ethylene gives terminal double bonds (Scheme 9).

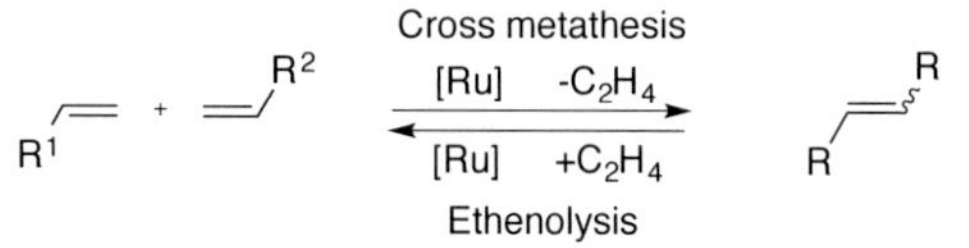

Scheme 9. General CM reaction.

Scheme 10 shows CM reactions starting from 3-buten-1-ol (**20**) including site product formation due to isomerization.

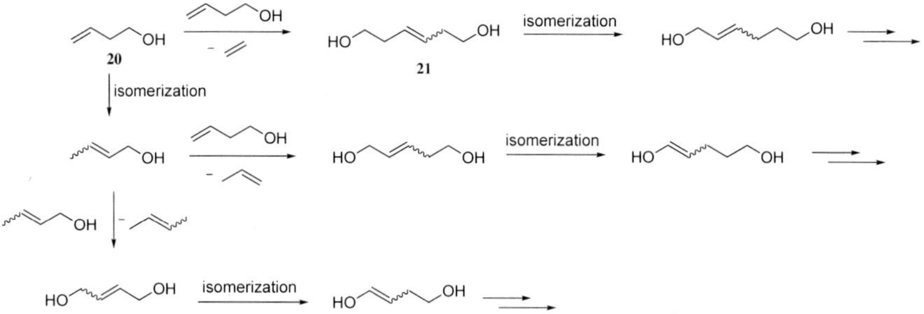

Scheme 10. CM and isomerization of **20**.

In this work, the influence of the protein **FhuA** on the catalyzed CM reaction of **20** was investigated. Because the protein free GH-type catalyst **COH** is not soluble in water, the reaction was performed in THF and compared to the protein catalyst **FhuA_C3**. All reactions were analyzed by GC. The conversion of **20** was monitored in each solvent (Figure 13).

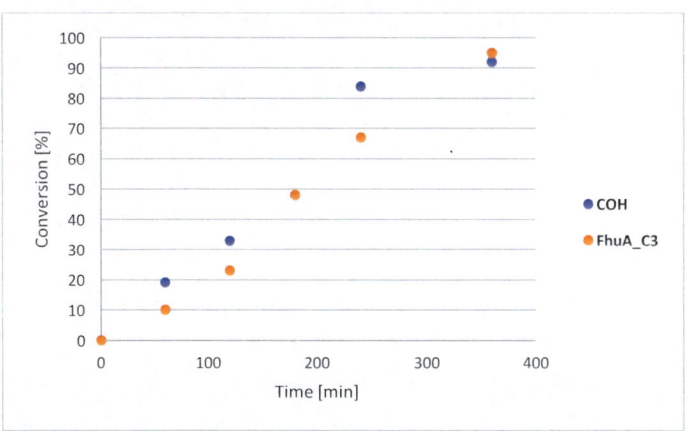

Figure 13. Time conversion plot of **20** with **COH** and **FhuA_C3**.

Despite the different solvents, both catalyst showed the same time-conversion profile. This allows to compare the catalysts and especially the selectivity over time. Because of the low solubility of **COH** in water, only a very slight conversion of **20** is observed after six hours if the suspension was vigorously stirred.

If the catalyst **COH** was employed in THF, a mixture of products was observed (Figure 14, top).

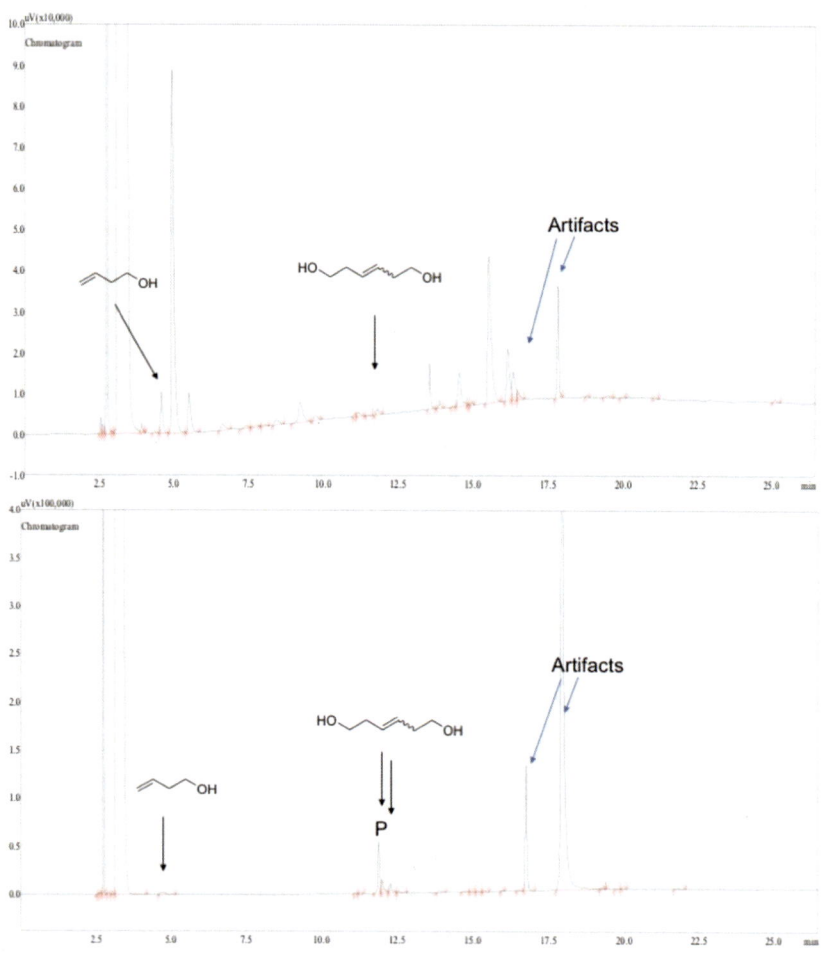

Figure 14. GC chromatograms of the CM of **20** catalyzed with **COH** (Top) and **FhuA_C3** (bottom). The artifacts observed are due to impurities derived from the GC machine.

In the GC chromatogram, the peak at 4.5 min corresponds to the starting material 3-buten-1-ol (**20**) (*cf.* Figure 16). Other peaks that results from isomerization (Scheme 10) may have been formed with the help of a catalyst that contains a [Ru-H] unit and which has been formed from decomposition of **COH**.[15-18] The formation of such a [Ru-H] fragment had been observed previously in alcohols or in aqueous media.[15,17-18]

The biohybrid catalyst **FhuA_C3** converted substrate **20** almost exclusively to 3-hexen-1,6-diol (**21**) (Figure 14, bottom). The protein probably alters the decomposition pathway. Isomerization does not occur without a [Ru-H] species. Similar findings in organic solvents were recently made by Mauduit and coworkers.[19] The group redesigned the NHC ligand of a GH-type catalyst and did not observe [Ru-H] in decomposition studies of the catalyst, yielding a clean CM reaction.[19] Another effect of the protein might be the increased diffusion barrier to the active site which prevents isomerization. A [Ru-H] species may also be deactivated by coordination to protein residues that would decrease the possible coordination sites at the metal center. The protein might also have a redox effect. Grubbs and coworkers reported that addition of 1,4-benzoquinone hampered the formation of [Ru-H] species and therefore decreased the isomerization significantly.[20] XAS studies may provide deeper insight into the decomposition of the catalyst within a protein.

B.1.5.4. Ethenolysis with FhuA Based Biohybrid Catalysts

Ethenolysis of oleic acid (**22**), fumaric acid (**23**) and maleic acid (**24**) was investigated (Figure 15).

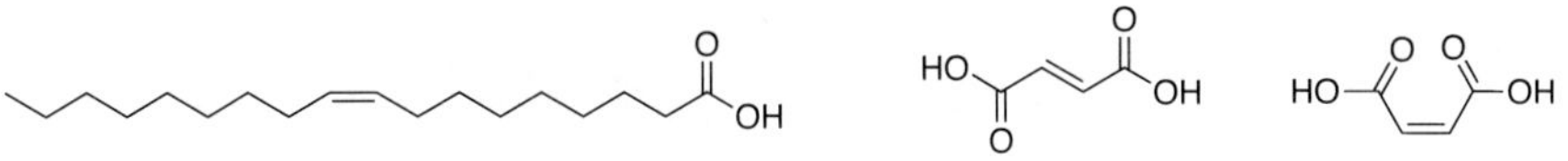

Figure 15. Oleic acid **22**, fumaric acid **23** and maleic acid **24**.

These reagents were treated with ethylene (1 bar) using the catalyst **COH** in aqueous media and in THF as well as with the catalyst **FhuA_C3** in aqueous SDS solution at 40 °C (Table 8).

Table 8. Ethenolysis of different substrates.

$$R\diagdown\diagup R \xrightarrow[\text{40 °C, 24 h, +C}_2\text{H}_4]{\text{catalyst (1 mol \%)}} R\diagdown\diagup + \diagup\diagdown R$$

Entry	Catalyst	Substrate	Solvent	Conv.[b] [%]
1	COH	23	THF-d_8	< 5
2	COH	23	THF-d_8/D_2O	0
3	FhuA_C3	23	H_2O	0
4	COH	22	THF-d_8	45
5	COH	22	THF-d_8/D_2O	0
6	FhuA_C3	22	H_2O	0
7	COH	24	THF-d_8	45
8	COH	24	THF-d_8/D_2O	0
9	FhuA_C3	24	H_2O	> 99

[a] c(substrate) = 0.05 M [b] Determined by ^{1}H NMR spectroscopy in $CDCl_3$ or GC.

The mild conditions required by the protein do not lead to cleavage of the *E* configured double bond of fumaric acid (**23**) (Table 8, entries 1-3). According to the literature, electron poor *E* configured double bonds usually require harsher reaction conditions.[21-22] The *Z* double bond of oleic acid is converted to 50% (Table 8, entry 4). This is due to the ratio between ethylene and oleic acid the NMR tube, where the ethylene is fully consumed at 50% conversion of **23**. In contrast, oleic acid in aqueous media is not converted neither by **COH** nor by **FhuA_C3** (Table 8, entries 5 and 6).

In the organic solvent THF, the double bond of maleic acid is cleaved and acrylic acid is obtained with ca. 50% conversion as with oleic acid. In aqueous media, maleic acid was not converted at all (Table 8, entry 8), but full conversion was achieved with the protein based catalyst **FhuA_C3**, where two products in a ration of 1:2 are obtained. The minor product was identified as lactic acid that was formed by hydrolysis of acrylic acid (anti-Markovnikov addition of water) that was generated by cleaving the C=C double bond in maleic acid *via* ethenolysis (^{1}H NMR spectrum after the reaction is shown in B.1.7. Figure 17). The other product was not identified (Scheme 11).

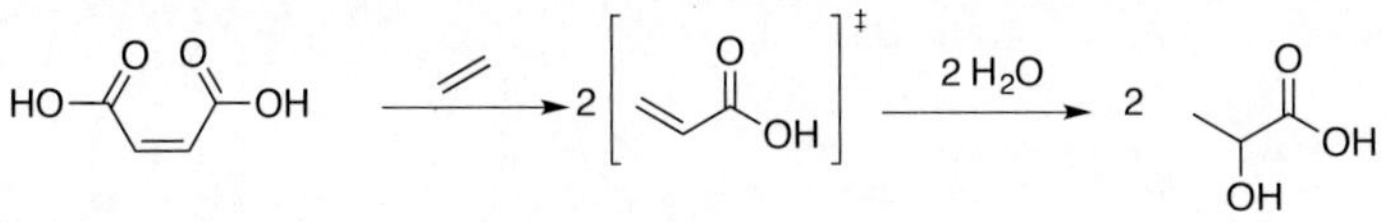

$$\text{HO}-\underset{\text{O}}{\overset{\text{O}}{\|}}-\underset{\text{O}}{\overset{\text{O}}{\|}}-\text{OH} \longrightarrow 2\left[\underset{\text{OH}}{\overset{\text{O}}{\|}}\right]^{\ddagger} \xrightarrow{2\,H_2O} 2\,\underset{\text{OH}}{\overset{\text{O}}{\|}}\text{OH}$$

Scheme 11. Proposed formation of lactic acid by **FhuA_C3**.

The Markovnikov hydration product 3-hydroxpropionic acid was not observed in this reaction. However, only this product was obtained if the acrylic acid was stirred under catalytic conditions without the protein catalyst. The catalyst **COH** in THF/aqueous SDS solution did not lead to any conversion. This may be due to micelles that separate catalyst and substrate.

B.1.6. Summary and Conclusion

GH-type catalysts **C1** and **C2** with a shorter spacer between the NHC ligand and the maleimide group have been synthesized. These were successfully conjugated to the transmembrane protein **FhuA**. Model compounds were formed by conjugating catalysts **C1-C3** to a cysteine. The labile ester bond complicated the observation of the metal containing fragment in MS analysis of the biohybrid conjugates as well as of the model compounds. Under MS conditions, the Ru-NHC bond or the ester bond at the backbone of the NHC are easily cleaved and the metal is released from the protein. ESI-TOF MS with very low ionization energy might help to detect the attached metal fragment, but a detailed protocol for the workup after the required TEV cleavage needs to be established.

The selectivity in the polymer **17** changed slightly with biohybrid catalysts **FhuA_C1-3**. The cysteine adducts **C1-C3_Cys** did not affect the selectivity compared to **C1-C3**. The selectivity can only be changed by a more complex environment of the catalytically active center what complicates the further development of these systems. The detergent in the folded state does not directly influence the catalyst, but the choice of detergent crucially affects the stability of the protein and allows to avoid the precipitation of the protein.

The biohybrid catalyst **FhuA_C3** allowed chemoselective transformations by cross metathesis. During CM of buten-4-ol (**20**), isomerization was suppressed and the formation of side products was not observed. No other artificial metathease that allows chemoselective cross-metathesis has been described. Furthermore, **FhuA_C3** catalyzed the ethenolysis reaction with maleic acid.

B.1.7. Experimental Section

All operations were performed under an inert atmosphere of argon or nitrogen using standard Schlenk or glove box techniques. Water was thoroughly degassed with argon or nitrogen prior to use (> 1 h). Other solvents were degassed by using "freeze-pump-thaw" technique. Diethyl ether, dichloromethane and tetrahydrofuran were obtained dry and degassed from a SPS 800 from *MBraun*. Dichloromethane-d_2 and chloroform-d_1 were dried over calcium hydride, distilled, degassed and stored in a glove box. THF-d_8 was dried over sodium/benzophenone, distilled, degassed and stored in a glove box. NMR-Spectra were recorded on a *Bruker* Avance III spectrometer (^{1}H, 400.1 MHz). Chemical shifts were referenced internally by using the residual solvent resonances.[23] CD spectra were recorded on a *JASCO* J-1100 equipped with a single position Peltier cell holder. MALDI–TOF MS spectra were recorded on an Ultraflex III TOF/ TOF mass spectrometer (*Bruker Daltonics*). High resolution ESI–TOF MS were performed on a Thermo Finnigan LCQ Deca XP Plus spectrometer. GC measurements were performed on a gas chromatograph from *Schimadzu* equipped with a capillary column and a FID. Compounds **C3**,[1] **FhuA_C3**,[1] diallylamine **3**,[24] 7-oxanorbornene derivative **16**[25-26] and diol **18**[27-28] were synthesized according to literature procedures. Cysteine **Cys**, buten-4-ol (**20**), oleic acid (**22**), fumaric acid (**23**) and maleic acid (**24**) are commercially available and were used as received. **FhuA_ΔCVF^{tev}** was prepared and analyzed as previously reported from the group of Prof. U. Schwaneberg at the Institute for Biotechnology at RWTH Aachen University. Digestion with a TEV protease was performed by the group of Prof. U. Schwaneberg at the Institute for Biotechnology at RWTH Aachen University.[1,9] All other chemicals were used as received if not mentioned otherwise.

General Procedure for the Conjugation of the GH-type catalyst to FhuA

To a degassed solution containing **FhuA_ΔCVF^{tev}** (5 mL, 5 mg/mL) and SDS (1% (w/w)), a GH-type catalyst **C1/C2/C3** (5 mg) in degassed THF (20% (v/v)) was added dropwise over a period of 10 min. The solution was stirred at 25 °C for 18 h. The mixture was lyophilized and washed with dry, degassed THF (3×15 mL). The residue was dried *in vacuo* and the greenish powder was dissolved in 5 mL degassed water.

General Procedure for the refolding of FhuA.

The aqueous solution of **FhuA_C1-3** containing SDS (1% (w/w)) was transferred into a dialysis tube (12-14 kDa MWCO). The tube was sealed and transferred into a beaker containing the refolding reagent in aqueous buffer solution (0.1 M NaP$_i$, pH = 7.4). For refolding, either MPD (50 mM) or PE-PEG (0.125 mM) was used. Buffer solution was exchanged at least three times (approximately every 12 h). Correct folding of the **FhuA** or the biohybrid conjugates was confirmed by CD spectroscopy after 48 h of dialysis. If the CD spectroscopy did not indicate the typical β-barrel shape, dialysis was continued for 24 h. Precipitates during refolding were removed by centrifugation.

General procedure for the ROMP of 7-oxanorbornene derivative 16.

A Schlenk tube was charged with the protein solution containing either the GH-type catalyst, **FhuA** or the biohybrid conjugate in the indicated buffer solution. To this solution, THF (10% (v/v)) was added and the solution stirred at room temperature for 15 minutes. Afterwards, substrate **16** (1000 equiv.) was added with a microsyringe. After reaction time, 100 equiv. of ethyl vinyl ether was added to quench the catalyst and cleave the polymer of the Ru.[29] The solvent was removed *in vacuo* and the residue was taken up in CDCl$_3$. The polymer was analyzed by ^{1}H NMR as described by Feast and Harrison.[26]

General Procedure for the RCM reaction

A Schlenk tube was charged with the protein solution containing either the GH-type catalyst, **FhuA** or the biohybrid conjugate in the indicated buffer solution. To this solution, THF (10% (v/v)) was added and the solution stirred at room temperature for 15 minutes. Afterwards, substrate **3** or **18** (100 or 1000 equiv.) was added with a microsyringe. The mixture was stirred at 40 °C under reduced pressure (ca. 750 mbar). After reaction time, 100 equiv. of ethyl vinyl ether was added to quench the catalyst.[29] The solvent was removed *in vacuo* and the residue was taken up in CDCl$_3$. The mixture was analyzed like reported previously.[3]

General Procedure for the CM reaction

A Schlenk tube was charged with the protein solution containing either the GH-catalyst, **FhuA** or the biohybrid conjugate in the indicated buffer solution. To this solution, THF (10% (v/v)) was added and the solution stirred at room temperature for 15 minutes. Afterwards, the corresponding substrate (100 equiv.) was added. After reaction time, 100 equiv. of ethyl vinyl ether was added to quench or an aliquot was taken and quenched by addition of ethyl vinyl ether.[29] The solvent was removed *in vacuo* and the residue was taken up in CDCl$_3$. The mixture was analyzed by ^{1}H NMR spectroscopy. In case of GC analysis, 500 μL THF were added and the sample was injected into the GC and analyzed.

Synthesis of dichloro-4-(((4-(2,5-dioxo-2,5-dihydro-1*H*-pyrrol-1-yl)ethanoyl)-oxy)-methyl)-1,3-dimesityl-4,5-dihydro-1H-imidazol-2-ylidene-(o-isopropoxy-phenylmethylene) ruthenium (IV) (C1)

In a glovebox in an oven dried schlenk flask, Grubbs-Hoveyda type complex **9** (30 mg, 0.05 mmol, 1.0 equiv.) was dissolved in dry THF (1 mL) and dry NaHCO$_3$ (30 mg, 0.36 mmol,

8 equiv.) was added. Acetyl chloride **10** (9 mg, 0.05 mmol, 1.1 equiv.) dissolved in THF (0.2 mL) was added dropwise to the solution. The mixture was stirred at 25 °C for 16 h. The solid was filtered off, and diethyl ether (25 mL) was added. The organic layer was washed with water (2×25 mL),brine (1×25 mL) and dried over MgSO$_4$. The solvent was removed *in vacuo* to yield complex **13** (35 mg, 0.05 mmol, 98%) as green powder. ^{1}**H NMR** (400 MHz, CD$_2$Cl$_2$, 23 °C): δ = 16.41 (s, 1H, Ru=C*H*), 7.58-7.54 (m, 1H, Ar-*H*), 7.10-7.08 (m, 4H, Ar-*H*), 6.97-6.89 (m, 2H, Ar-*H*), 6.85 (d, $^3J_{H,H}$ = 8.1 Hz, 1H, Ar-*H*), 6.77 (s, 2H, C*H*=C*H*), 4.90 (sept, *J* = 6.1 Hz, 1H, C*H*(CH$_3$)$_2$), 4.70-4.63 (m, 1H, C*H*), 4.37-4.18 (br. m, 5H, C*H*, C*H*$_2$, NC*H*$_2$COO), 4.01-3.96 (m, 1H, C*H*$_2$), 2.43 (br. s, 18H, Ar-*Me*), 1.27-1.23 (m, 6H, CH(C*H*$_3$)$_2$) ppm. ^{13}**C-NMR** (100 MHz; CD$_2$Cl$_2$; 25 °C): δ = 297.03, 214.54, 170.34, 152.59, 145.61, 139.56, 139.52, 135.03, 130.59, 130.27, 130.22, 130.03, 129.86, 122.89, 122.78, 113.51, 75.69, 66.25, 66.02, 63.30 (br), 38.77, 30.65, 30.24, 21.55, 21.38, 21.31, 19.78 (br), 15.57 ppm. **ESI-HR MS** (positive mode): m/z calcd. for C$_{38}$H$_{43}$Cl$_2$N$_3$O$_5$RuNa [M+Na]$^+$: 816.1519; Found: 816.1464.

Synthesis of dichloro-4-(((4-(2,5-dioxo-2,5-dihydro-1*H*-pyrrol-1-yl)propanoyl)-oxy)-methyl)-1,3-dimesityl-4,5-dihydro-1H-imidazol-2-ylidene-(o-isopropoxy-phenylmethylene) ruthenium (IV) (C2)

In a glovebox a schlenk flaskwas charged with a Grubbs-Hoveyda type complex **9** (30 mg, 0.05 mmol, 1.0 equiv.) dissolved in dry THF (1 mL) and dry NaHCO$_3$ (30 mg, 0.36 mmol, 8 equiv.) was added. Acetyl chloride **11** (9 mg, 0.05 mmol, 1.1 equiv.) in THF (0.2 mL) was added dropwise to the solution. The mixture was allowed to stir at 25 °C for 16 h. The solid was filtered off, and diethyl ether (25 mL) was added. The organic layer was washed with water (2×25 mL) and brine (1×25 mL). The organic layer was dried over MgSO$_4$, and the solvent removed in vacuum to yield complexes **14** (36 mg, 0.05 mmol, 98%) as green powder. ^{1}**H NMR** (400 MHz, CD$_2$Cl$_2$, 23 °C): δ = 16.48 (s, 1H, Ru=C*H*), 7.62-7.56 (m, 1H, Ar-*H*), 7.13-7.11 (m, 4H, Ar-*H*), 7.01-6.93 (m, 2H, Ar-*H*), 6.88 (d, $^3J_{H,H}$ = 8.2 Hz, 1H, Ar-*H*), 6.75 (s, 2H, C*H*=C*H*), 4.94 (sept, $^3J_{H,H}$ = 6.1 Hz, 1H, C*H*(CH$_3$)$_2$), 4.52-4.42 (m, 1H, C*H*), 4.33 (ψt, $^3J_{H,H}$ = 10.4 Hz, 1H, C*H*$_2$), 4.14-4.09 (m, 2H, C*H*$_2$), 3.86 (t, $^3J_{H,H}$ = 7.0 Hz, 2H, C*H*$_2$) 3.87-3.78 (m, 1H, C*H*$_2$), 2.85 (t, $^3J_{H,H}$ = 7.0 Hz, 2H, C*H*$_2$), 2.47 (br. s, 18H, Ar-*Me*), 1.30-1.22 (m, 6H, CH(C*H*$_3$)$_2$) ppm. ^{13}**C NMR** (100 MHz; CD$_2$Cl$_2$; 25 °C): δ = 296.91, 214.48, 170.92, 168.71, 152.57, 145.60, 139.68, 139.50 (br), 134.81, 130.58, 130.26, 130.19, 130.02, 129.84, 122.88, 122.77, 113.50, 75.68, 66.06 (br), 63.33 (br), 34.31, 33.33, 23.73, 21.54, 21.37, 21.29, 19.36 (br) ppm. **ESI-HR MS** (positive mode): m/z calcd. for C$_{39}$H$_{45}$Cl$_2$N$_3$O$_5$RuNa [M+Na]$^+$: 830.1676; Found: 830.1663.

Synthesis of the Cysteine Adduct of C1 (C1_Cys)

Catalyst **C1** (30.0 mg, 0.037 mmol, 1.00 equiv.) was dissolved in THF (1 mL). L-Cysteine (9.00 mg, 0.04 mmol, 1.00 equiv.) was dissolved in THF (1 mL) and added dropwise to the solution of **C1** in THF. To this mixture, NMM (1 µL) was added. The solution was stirred for 4 h at 25 °C. The solvent was removed in vacuum and **C1_Cys** was obtained as greenish solid (36 mg, 0.04 mmol, 99%).

^{1}**H NMR** (400, MHz, THF, 23 °C): δ = 16.28 (s, 1H, Ru=CH), 7.60-7.52 (m, 1H, Ar-H), 7.15-7.05 (m, 4H, Ar-H), 6.95-6.91 (m, 2H, Ar-H), 6.82 (d, J = 7.34 Hz, 1H, Ar-H), 4.91 (sept, J = 6.24 Hz, 1H, CH(CH)$_3$), 4.44-4.39 (m, 1H, CH), 4.32 (m, 1H, CH_2), 3.87-3.80 (m, 2H, NCH_2), 3.79-3.73 (m, 1H, CH_2), 3.68 (s, 3H, OCH_3), 3.25-3.12 (m, 2H, CH_2CHS), 2.90-2.82 (m, 2H, CH_2COO), 2.81-2.70 (m, 2H, CHSCH_2) 2.50-2.30 (br m, 18H, CH_3), 1.42 (s, 9H, O(CH_3)$_3$), 1.30-1.28 (m, 6H, CH(CH_3)$_2$) ppm.

Synthesis of the Cysteine Adduct of C2 (C2_Cys)

Catalyst **C2** (30.0 mg, 0.037 mmol, 1.00 equiv.) was dissolved in THF (1 mL). L-Cysteine (9.00 mg, 0.04 mmol, 1.00 equiv.) was dissolved in THF (1 mL) and added dropwise to the solution of **C1** in THF. To this mixture, NMM (1 µL) was added. The solution was stirred for 4 h at 25 °C. The solvent was removed in vacuum and **C2_Cys** was obtained as greenish solid (36 mg, 0.04 mmol, 99%).

^{1}**H NMR** (400 MHz, CD$_2$Cl$_2$, 23 °C): δ = 16.44 (s, 1H, Ru=CH), 7.57-7.51 (m, 1H, Ar-H), 7.12-7.04 (m, 4H, Ar-H), 6.94-6.90 (m, 2H, Ar-H), 6.85 (d, J = 8.25 Hz, 1H, Ar-H), 4.91 (sept, J = 6.24 Hz, 1H, CH(CH)$_3$), 4.48-4.37 (m, 1H, CH), 4.31 (ψt, J = 10.45 Hz, 1H, CH_2), 4.13-4.07 (m, 2H, CH_2), 3.86-3.80 (m, 2H, NCH_2), 3.79-3.76 (m, 1H, CH_2), 3.75 (s, 3H, OCH_3), 3.19-3.09 (m, 2H, CH_2CHS), 2.94 (t, J = 7.11 Hz, 2H, CH_2COO), 2.91-2.86 (m, 2H, CHSCH_2) 2.50-2.30 (br m, 18H, CH_3), 1.44 (s, 9H, O(CH_3)$_3$), 1.31-1.25 (m, 6H, CH(CH_3)$_2$) ppm.

Synthesis of the Cysteine Adduct of C3 (C3_Cys)

Catalyst **C3** (30.0 mg, 0.037 mmol, 1.00 equiv.) was dissolved in THF (1 mL). L-Cysteine (9.00 mg, 0.04 mmol, 1.00 equiv.) was dissolved in THF (1 mL) and added dropwise to the solution of **C1** in THF. To this mixture, NMM (1 µL) was added. The solution was stirred for 4 h at 25 °C. The solvent was removed in vacuum and **C3_Cys** was obtained as greenish solid (36 mg, 0.04 mmol, 99%).

^{1}**H NMR** (400 MHz, CD$_2$Cl$_2$, 23 °C): δ = 16.44 (s, 1H, Ru=CH), 7.57-7.51 (m, 1H, Ar-H), 7.12-7.04 (m, 4H, Ar-H), 6.94-6.90 (m, 2H, Ar-H), 6.85 (d, J = 8.25 Hz, 1H, Ar-H), 4.91 (sept, J = 6.24 Hz, 1H, CH(CH)$_3$), 4.48-4.37 (m, 1H, CH), 4.31 (ψt, J = 10.45 Hz, 1H, CH_2), 4.13-4.07 (m, 2H, CH_2), 3.86-3.80 (m, 2H, NCH_2), 3.79-3.76 (m, 1H, CH_2), 3.75 (s, 3H, OCH_3),

3.19-3.09 (m, 2H, CH_2CHS), 2.94 (t, J = 7.11 Hz, 2H, CH_2COO), 2.91-2.86 (m, 2H, CHSCH_2) 2.50-2.30 (br m, 18H, CH_3), 1.89 (m, 2H, CH_2CH_2CH_2), 1.44 (s, 9H, O(CH_3)$_3$), 1.31-1.25 (m, 6H, CH(CH_3)$_2$) ppm.

ESI-TOF MS spectra of C2 and C2_Cys.

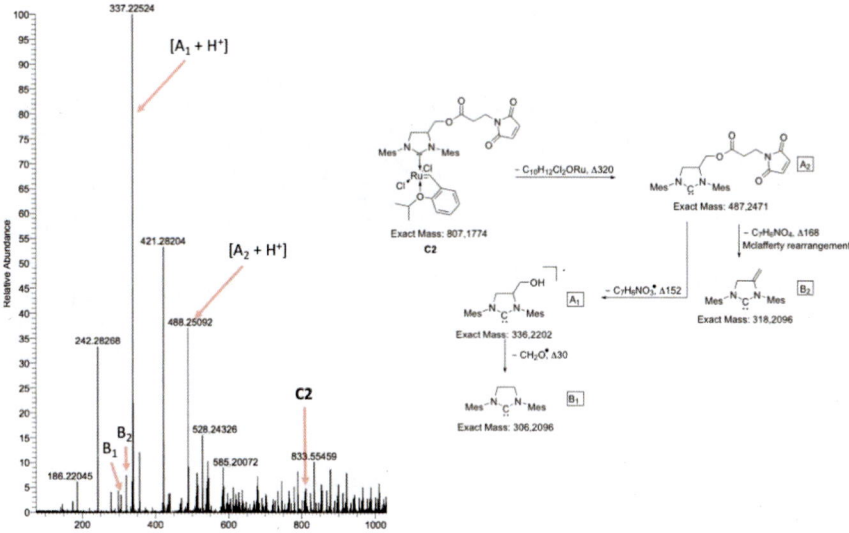

Scheme 12. ESI spectrum and possible fragmentation of **C2**.

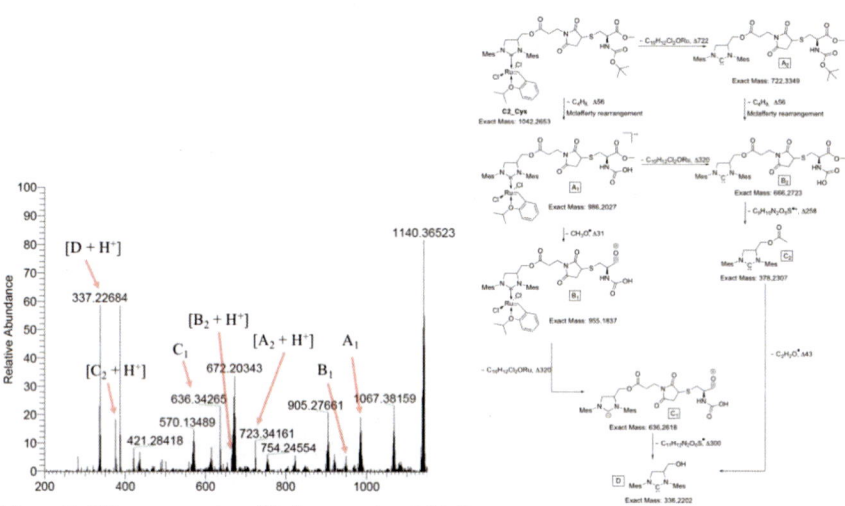

Scheme 13. ESI spectrum and possible fragmentation of **C2_Cys**.

41

Full list of ROMP reactions
Table 9. ROMP of **16** in THF.

Entry[a]	Catalyst	Conv.[b] [%]	cis/trans[b]	TON
1	COH	> 99	71/29	> 990
2	C1	> 99	70/30	> 990
3	C2	> 99	72/28	> 990
4	C3	> 99	69/31	> 990
5	C1_Cys	> 99	70/30	> 990
6	C2_Cys	> 99	72/28	> 990
7	C3_Cys	> 99	71/29	> 990

[a] $c(\mathbf{16}) = 0.1$ M [b] Determined by ^{1}H NMR spectroscopy in $CDCl_3$.

Table 10. ROMP of **16** in aqueous SDS solution.

Entry[a]	Catalyst	Conv.[b] [%]	pH	cis/trans[b]	TON[c]
1	COH	> 99	5.8	60/40	> 990
2	COH	85	7.4	59/41	850
3	C1	> 99	5.8	60/40	> 990
4	C1	90	7.4	60/40	900
5	C2	> 99	5.8	61/39	> 990
6	C2	89	7.4	59/41	890
7	C3	99	5.8	60/40	990
8	C3	91	7.4	60/40	910
9	C1_Cys	> 99	5.8	60/40	> 990
10	C1_Cys	89	7.4	59/41	890
11	C2_Cys	99	5.8	61/39	990
12	C2_Cys	90	7.4	60/40	900
13	C3_Cys	99	5.8	60/40	990
14	C3_Cys	91	7.4	61/39	910
15	FhuA_C1*	32	5.8	63/37	320
16	FhuA_C1*	63	7.4	59/41	630
17	FhuA_C2*	77	5.8	50/50	770
18	FhuA_C2*	27	7.4	50/50	270

[a] $c(\mathbf{16})$ = 0.1 M, THF 10% (v/v). [b] Determined by ^{1}H NMR spectroscopy in CDCl$_3$. [c] Based on the results using ThioGlo titration and BCA assay.

Table 11. ROMP of **16** in aqueous THF solution.

Entry[a]	Catalyst	Conv.[b] [%]	pH	cis/trans[b]	TON
1	COH	> 99	5.8	50/50	> 990
2	COH	99	7.4	50/50	990
5	C1	> 99	5.8	50/50	> 990
6	C1	98	7.4	51/49	980
7	C2	> 99	5.8	50/50	> 990
8	C2	99	7.4	50/50	990
9	C3	> 99	5.8	50/50	> 990
10	C3	99	7.4	50/50	990
11	C1_Cys	> 99	5.8	51/49	> 990
12	C1_Cys	99	7.4	50/50	990
13	C2_Cys	> 99	5.8	50/50	> 990
14	C2_Cys	99	7.4	50/50	990
15	C3_Cys	> 99	5.8	50/50	> 990
16	C3_Cys	99	7.4	51/49	990

[a] $c(\mathbf{16})$ = 0.1 M, THF 10% (v/v). [b] Determined by ^{1}H NMR spectroscopy in CDCl$_3$.

GC Chromatogram

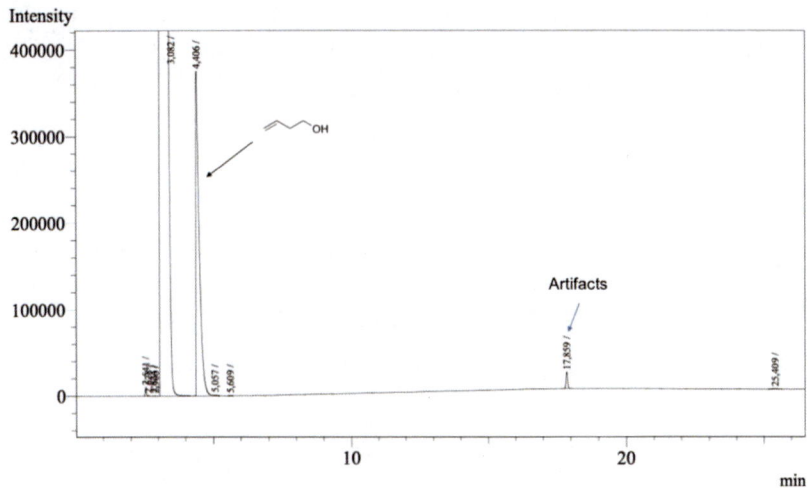

Figure 16. GC chromatogram of 1-buten-4-ol 20.

^{1}H NMR spectrum of product mixture derived from ethenolysis of maleic acid 23 with biohybrid conjugate.

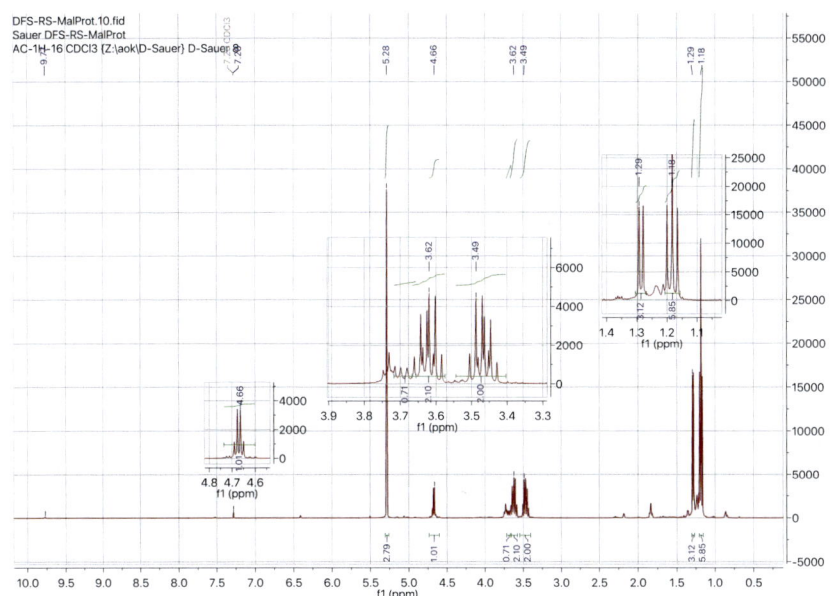

Figure 17. ^{1}H NMR spectra (CDCl$_3$, 23 °C) after ethenolysis of 24 with bioyhbrid catalyst **FhuA_C3**.

B.1.8. References

(1) Philippart, F.; Arlt, M.; Gotzen, S.; Tenne, S.-J.; Bocola, M.; Chen, H.-H.; Zhu, L.; Schwaneberg, U.; Okuda, J. *Chem. Eur. J.* **2013**, *19*, 13865.

(2) Tenne, S.-J.; Schwaneberg, U. *Int. J. Mol. Sci.* **2012**, *13*, 2459.

(3) Gotzen, S. Doctoral Dissertation, RWTH Aachen University, **2016**.

(4) Hermanson, G. T. In *Bioconjugate Techniques (Third edition)*; Academic Press: Boston, **2013**, p 229.

(5) de Figueiredo, R. M.; Oczipka, P.; Fröhlich, R.; Christmann, M. *Synthesis* **2008**, 1316.

(6) Rahman, A.; Brown, C. W. *J. Appl. Polym. Sci.* **1983**, *28*, 1331.

(7) Li, L.; Tan, Y. B. *J. Colloid Interface Sci.* **2008**, *317*, 326.

(8) Sauer, D. F.; Bocola, M.; Broglia, C.; Arlt, M.; Zhu, L.-L.; Brocker, M.; Schwaneberg, U.; Okuda, J. *Chem. Asian J.* **2015**, *10*, 177.

(9) Zhu, L.; Arlt, M.; Liu, H.; Bocola, M.; Sauer, D. F.; Gotzen, S.; Okuda, J.; Schwaneberg, U. In *Bio-Synthetic Hybrid Materials and Bionanoparticles: A Biological Chemical Approach Towards Material Science*; The Royal Society of Chemistry: **2015**, p 57.

(10) Lo, C.; Ringenberg, M. R.; Gnandt, D.; Wilson, Y.; Ward, T. R. *Chem. Commun.* **2011**, *47*, 12065.

(11) Mayer, C.; Gillingham, D. G.; Ward, T. R.; Hilvert, D. *Chem. Commun.* **2011**, *47*, 12068.

(12) Matsuo, T.; Imai, C.; Yoshida, T.; Saito, T.; Hayashi, T.; Hirota, S. *Chem. Commun.* **2012**, *48*, 1662.

(13) Basauri-Molina, M.; Verhoeven, D. G. A.; van Schaik, A. J.; Kleijn, H.; Klein Gebbink, R. J. M. *Chem. Eur. J.* **2015**, *21*, 15676.

(14) Zhao, J.; Kajetanowicz, A.; Ward, T. R. *Org. Biomol. Chem.* **2015**, *13*, 5652.

(15) Tomasek, J.; Schatz, J. *Green Chem.* **2013**, *15*, 2317.

(16) Dinger, M. B.; Mol, J. C. *Organometallics* **2003**, *22*, 1089.

(17) Beach, N. J.; Lummiss, J. A. M.; Bates, J. M.; Fogg, D. E. *Organometallics* **2012**, *31*, 2349.

(18) Alcaide, B.; Almendros, P.; Luna, A. *Chem. Rev.* **2009**, *109*, 3817.

(19) Rouen, M.; Queval, P.; Borré, E.; Falivene, L.; Poater, A.; Berthod, M.; Hugues, F.; Cavallo, L.; Baslé, O.; Olivier-Bourbigou, H.; Mauduit, M. *ACS Catal.* **2016**, *6*, 7970.

(20) Hong, S. H.; Sanders, D. P.; Lee, C. W.; Grubbs, R. H. *J. Am. Chem. Soc.* **2005**, *127*, 17160.

(21) Bidange, J.; Fischmeister, C.; Bruneau, C. *Chem. Eur. J.* **2016**, *22*, 12226.

(22) Marx, V. M.; Herbert, M. B.; Keitz, B. K.; Grubbs, R. H. *J. Am. Chem. Soc.* **2013**, *135*, 94.
(23) Fulmer, G. R.; Miller, A. J. M.; Sherden, N. H.; Gottlieb, H. E.; Nudelman, A.; Stoltz, B. M.; Bercaw, J. E.; Goldberg, K. I. *Organometallics* **2010**, *29*, 2176.
(24) Che, C.; Li, W.; Lin, S.; Chen, J.; Zheng, J.; Wu, J.-c.; Zheng, Q.; Zhang, G.; Yang, Z.; Jiang, B. *Chem. Commun.* **2009**, 5990.
(25) Novak, B. M.; Grubbs, R. H. *J. Am. Chem. Soc.* **1988**, *110*, 960.
(26) Feast, W. J.; Harrison, D. B. *Polym. Bull.* **1991**, *25*, 343.
(27) Bouhadir, K. H.; Zhou, J. L.; Shevlin, P. B. *Synth. Commun.* **2005**, *35*, 1003.
(28) Adrio, L. A.; Quek, L. S.; Taylor, J. G.; Kuok Hii, K. *Tetrahedron* **2009**, *65*, 10334.
(29) Vougioukalakis, G. C.; Grubbs, R. H. *Chem. Rev.* **2009**, *110*, 1746.

B.2. Metathease Based on the Soluble Protein Nitrobindin[a]

B.2.1. Introduction

The heme protein nitrobindin (**NB**) belongs to the group of β-barrel proteins.[1] This small β-barrel protein consists of ten β-strands. Hayashi and coworkers were the first to use this protein as host for metal complexes.[2-4] The heme ligating histidine residues were removed by mutations to generates an empty cavity. Position 96 was mutated to a cysteine to allow covalent anchoring of a (cyclopentadienyl)rhodium complex via a maleimide group with an ethylene spacer at the cyclopentadienyl moiety ring (Figure 18).[2-3]

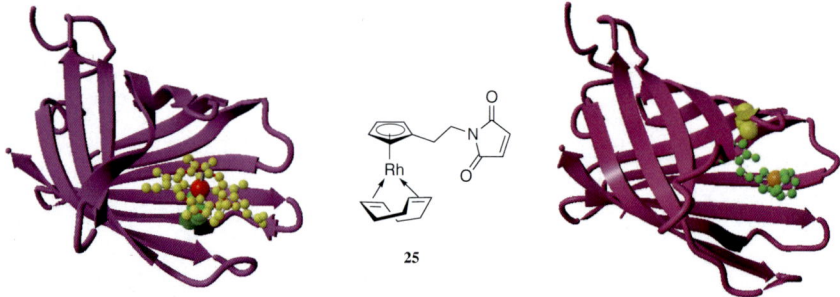

Figure 18. Left: Native **NB** containing heme unit (porphyrin: yellow, Fe: red, PDB: 3emm);[1] Middle: Rhodium catalyst (**25**) used by Hayashi and coworkers;[2] Right: Crystal structure of [**Rh**] attached to **NB4** (PDB: 3wjc).[3]

According to ICP-AES up to one metal was attached per protein.[3] Covalent anchoring was indicated by the molecular mass from MALDI-TOF MS that agrees with the calculated value.[3] Most **NB** apoproteins were crystallographically characterized. A crystal structure was also obtained for a protein mutant containing the [Rh] metal cofactor (Figure 18, right side).[3]

The rhodium based biohybrid conjugate was used in the polymerization reaction of phenylacetylene **33** where the *cis/trans* ratio of the double bond within the polymer chain was investigated (Table 12).[2-3]

---

[a] Parts of this chapter have been published in: Sauer, D. F.; Himiyama, T.; Tachikawa, K.; Fukumoto, K.; Onoda, A.; Mizohata, E.; Inoue, T.; Bocola, M.; Schwaneberg, U.; Hayashi, T.; Okuda, J. *ACS Catal.* **2015**, *5*, 7519.

Table 12. Polymerization of **33** catalyzed by biohybrid catalysts.

Entry	Catalyst[a]	cis/trans[b]	Cavity Volume [Å^3]	M_n[c]	PDI[c]
1	Mb(A125C)-Rh	91/9	surface mutant	46,500	2.3
2	NB(Q96C)-Rh	46/54	n.d.	42,800	2.1
3	NB1-Rh	56/44	1161	36,500	2.6
4	NB2-Rh	55/45	1006	38,300	2.2
5	NB3-Rh	45/55	1145	36,800	2.2
6	NB4-Rh	18/82	855	38,900	2.4
7	NB5-Rh	23/77	850	38,300	2.3
8	NB6-Rh	61/39	891	38,300	2.4
9	NB7-Rh	66/34	827	31,100	2.5
10	NB8-Rh	78/22	889	31,500	2.4
11	NB9-Rh	60/40	651	34,600	2.1
12	NB10-Rh	78/22	665	38,600	2.7

[a] [Rh] = 25. [b] Determined by ^{1}H NMR spectroscopy. [c] Determined by GPC in CHCl$_3$.

Ten mutants with different cavity volume and surrounding of the active [Rh] site were investigated.[2-3] A myoglobin mutant **Mb(A125C)** with a cysteine on the surface was chosen as reference. This protein does not provide a defined second coordination sphere for the metal site. This resulted in a polymer with exclusive *cis*-selectivity as obtained by polymerization in organic solvents (Table 12, entry 1). The mutants with larger cavity size (cavity volume > 1000 Å^3) gave a *cis/trans* ratio of ca. 50/50 (Table 12, entries 2-5). Narrower cavities favored the formation of *trans* products (Table 12, entries 6 and 7). The *cis* product was obtained for catalysts providing huge steric bulk to the metal site (Table 12, entries 8-12). Yield, molecular weight and PDI did not significantly depend on the mutation. This study shows, that the selectivity can be manipulated by the second coordination sphere which may be modified by mutations. The Hayashi and coworkers also investigated the Diels-Alder reaction catalyzed by a copper terpy complex (Scheme 14).[5]

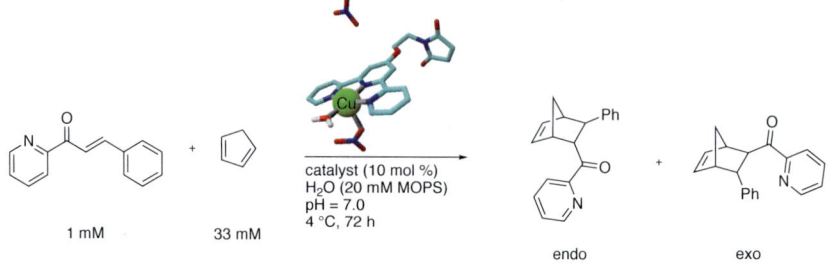

Scheme 14. Diels-Alder reaction performed by Hayashi and coworkers.[5]

NB as a protein host slightly changed the *endo/exo* selectivity and doubled the conversion in this reaction (Scheme 14). The increased conversion was ascribed to the accumulation of the hydrophobic substrates within the hydrophobic cavity.[5]

Another study by Lewis and coworkers used this protein to host a manganese terpy complex for olefin epoxidation reactions.[6] **NB** is particularly suited to construct biohybrid catalysts. It is relatively robust towards organic co-solvents and is easily modifiable to influence the active site.[2-6]

In this chapter, the influence of the hydrophobic cavity provided by **NB** on GH-type catalysts will be investigated. The results will be compared to the reported protein hosts for artificial metatheases (MjSHP,[7] **FhuA**,[8-10] (strept)avidin,[11-13] lipase,[14] hCAII[15] and α-chymotrypsin[16]) containing a hydrophilic cavity.

B.2.2. Synthesis and Characterization of Biohybrid Conjugates Based on NB

NB4 (mutations compared to WT_NB: M75L/H76L/Q96C/M148L/H158L) was used as host for the GH-type catalysts C1-C3 because this protein drastically influences the selectivity in the polymerization of phenylacetylene.[3] However, the small cavity did not allow the proper incorporation of the catalysts. A coupling was not observed with the shorter spacer of C1 and C2; the longer spacer of C3 allowed 25% coupling efficiency. The catalyst volumes from 752 Å^3 to 795 Å^3 (C1→C3) are still too big for the small cavity of 855 Å^3 in NB4. Additional steric bulk from the mesitylene groups of the NHC ligand further restricts the flexible movement of the catalyst. As a mutant with large cavity size, NB1 (mutations compared to WT_NB: L75A/H76L/Q96C/ M148L/H158A, 1161 Å^3 cavity size) was chosen (Scheme 15).

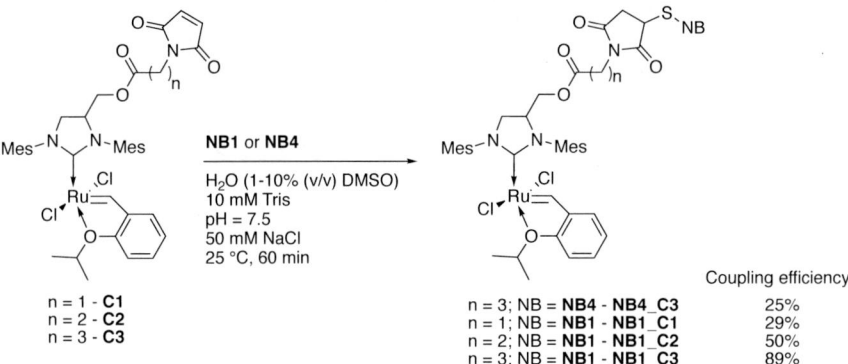

Scheme 15. Coupling of GH-type catalysts to variants of NB.

This increased the coupling efficiency (Scheme 15) that was determined from the ratio of the metal content (from ICP-AES) to the protein concentration (UV/vis spectroscopy after calibration with samples whose concentrations were determined by the Bradford assay). Figure 19 shows the UV/vis spectra of the conjugates and of the apoprotein.

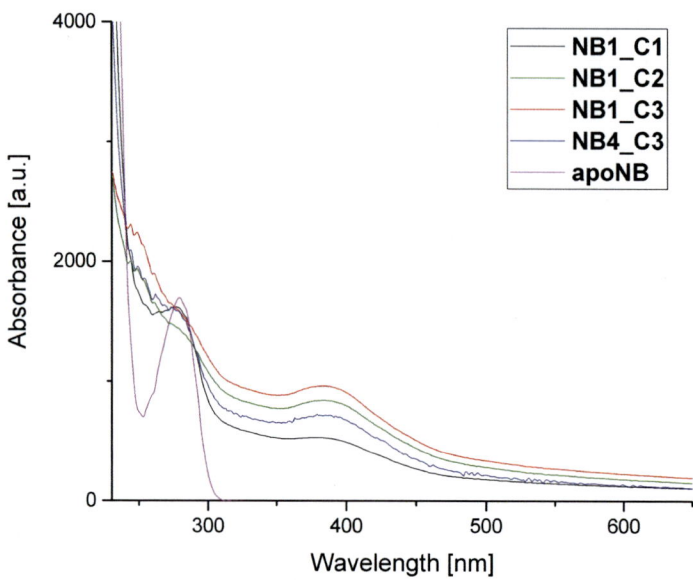

Figure 19. UV/vis spectra of **NB** and of the corresponding conjugates.

The UV/vis spectra indicate successful anchoring of the GH-type catalyst to the protein. The apoproteins **NB4** and **NB1** show an absorbance at 280 nm due to the aromatic amino acid residues in the protein. The conjugates show a second absorption at 380 nm, resulting from the metal to ligand charge transfer (MLCT) of the GH-type catalyst.[16]

CD spectroscopy confirmed structural integrity of the protein before and after anchoring. The spectra of **NB1** and **NB4** as well as for all biohybrid conjugates show a similar shape. Therefore, only the spectra for **NB1** and the **NB1_C3** catalyst at various temperatures are shown in Figure 20.

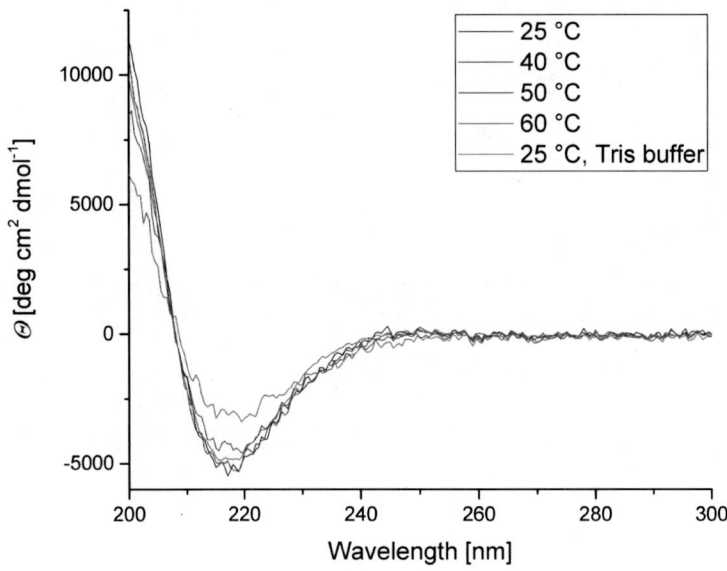

Figure 20. CD spectra of **NB1_C3** in H_2O (5 mM MES, pH = 6.0, 200 mM NaCl).

The CD spectrum for typical β-barrel proteins shows a minimum between 215-220 nm and a maximum around 195 nm.[17] These can be observed in the CD spectrum for **NB1** either in Tris buffer (Figure 20, green curve) or in MES buffer (Figure 20, black curve) at 25 °C. The β-barrel structure should be retained if the coupling is performed in Tris buffer or if MES buffer is employed during catalysis. CD spectra at varied temperature (Figure 20, red, blue and magenta curve) show stability up to 50 °C. At temperatures higher than 60 °C, precipitation of the protein is observed.

The MALDI-TOF MS spectra show only distribution of broad signals (Figure 21). This might be explained by the formation of products from a metathesis reaction between the catalyst and the double bond within the sinapic acid matrix employed for the measurement. However, the results of ESI-TOF MS performed with **NB1_C3** clearly indicate covalent anchoring (Figure 18).

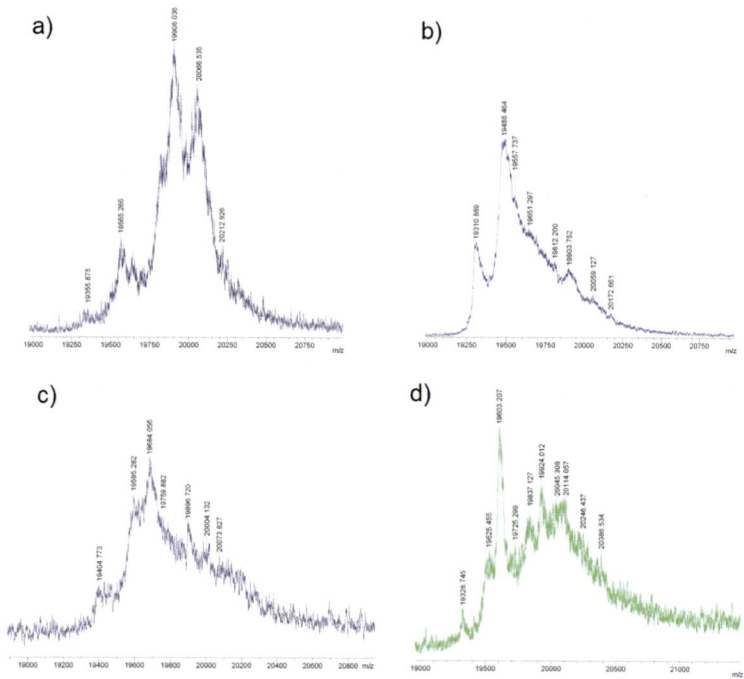

Figure 21. MALDI-TOF MS spectra of a) **NB4_C3** (calculated: 20238.71), b) **NB1_C1** (20126.19), c) **NB1_C2** (20154.55), and d) **NB1_C3** (20154.55). Calculated mass of the apoproteins: **NB4**: 19416.92; **NB1**: 19332.76. The broad peak distribution occurs due to reaction of the catalyst with sinapic acid used as a matrix.

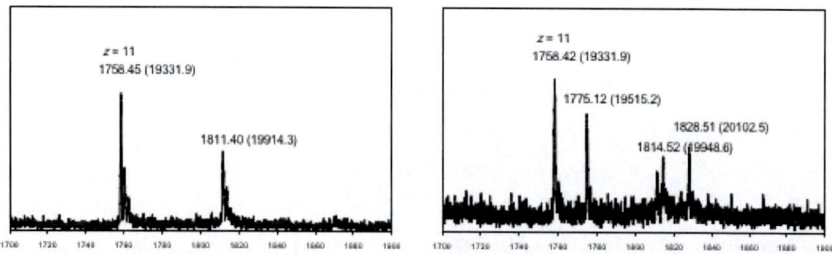

Figure 22. ESI-TOF MS spectra of **NB1** (left) and **NB1_C3** (right). The numbers in parentheses indicate the deconvoluted masses. Apo-**NB1**: 19331.6 (calculated: 19332.8); **NB1** + C8H9NO4: 19515.3 (ester cleavage; calculated: 19515.8); **NB1** + **COH** -Cl: 19945.7 (catalyst fragment of ester cleavage + **NB1**; calculated: 19952.4); **NB1_C3** -2Cl +H_2O: 20102.6 (calculated: 20102.0).

As a control experiment to verify the purification protocol used for the biohybrid conjugates, apo-**NB1** was stirred with **COH** under coupling conditions. After workup using a

HiTrap desalting column, trace metal analysis of the sample by ICP-AES did not show any ruthenium. This proves that the purification method is suitable for the biohybrid catalysts.

The larger volume of the cavity in **NB1** is needed to incorporate the bulky GH-type catalyst. However, this cavity still seems too small to incorporate more bulky catalysts and those with shorter spacer. Limited possibilities to mutate further single residues within the cavity on **NB1** call for a new approach to create a larger space within the cavity.

B.2.3. Synthesis and Characterization of Biohybrid Conjugates Based on NB4exp

B.2.3.1. Characterization of apo-NB4exp

In the group of Prof. Schwaneberg a strategy to expand the barrel structure was applied to **NB4** by M.Sc. Alexander R. Grimm. Instead of subjecting single residues to mutagenesis, two β-strands were copied, to create the new nitrobindin mutant **NB4exp** (Figure 23).

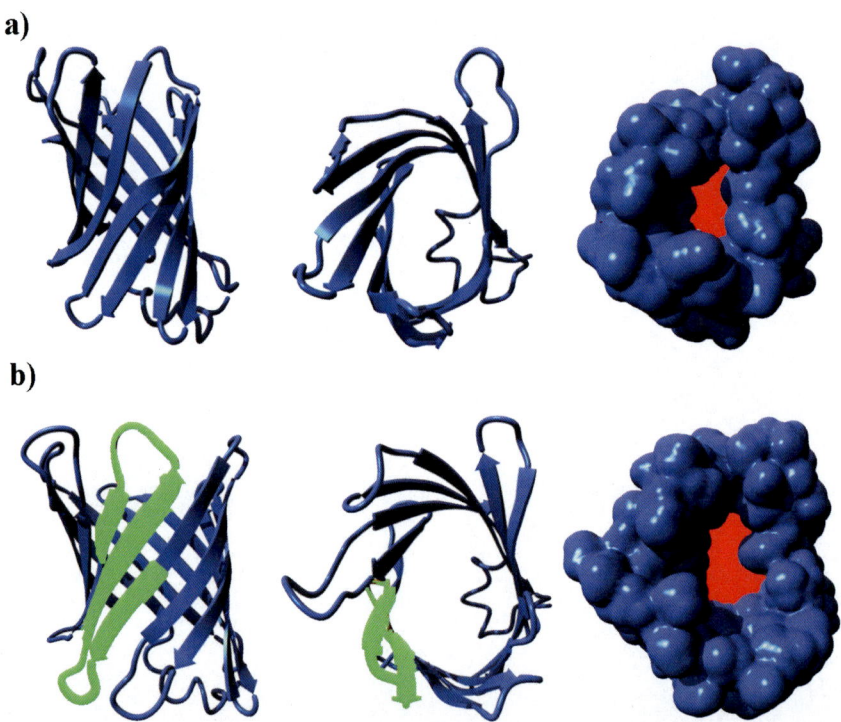

Figure 23. a) **NB4**: side view (cartoon), top view (cartoon) and top view (solvent accessible surface) on the cavity. b) **NB4exp**: side view (cartoon), top view (cartoon) and top view (solvent accessible surface) on the cavity. The copied β-sheets are shown in green.

This allowed to expand the barrel structure from 10 to 12 β-strands and to increase the cavity volume from 855 Å^3 in **NB4** to approximately 1400 Å^3 in **NB4exp**. CD spectroscopy revealed correct folding of the **NB4exp** mutant. Both proteins were investigated by VT_CD spectroscopy (Figure 24).

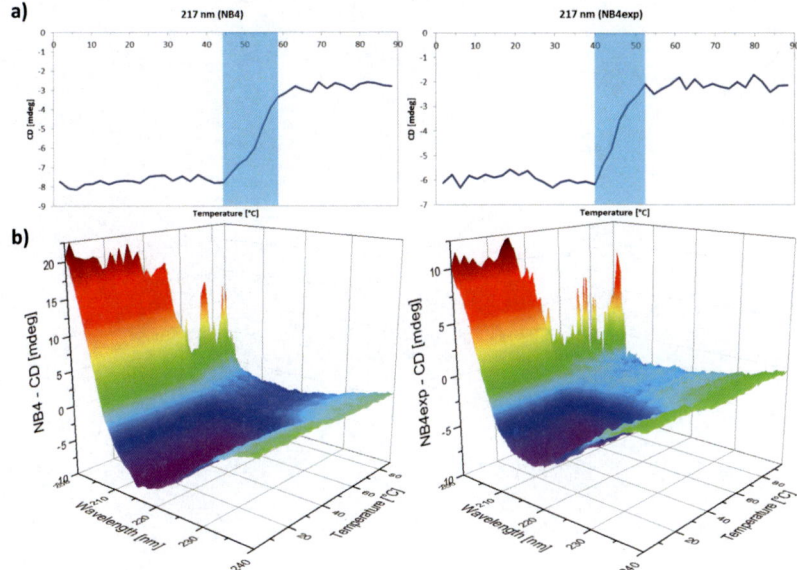

Figure 24. VT_CD spectra of **NB4** and **NB4exp**.

The **NB4** mutant is stable up to ca. 50 °C. In contrast, **NB4exp** shows an earlier melting at 45 °C. Both proteins precipitate at temperatures above 60 °C. The different in temperature stability is explained by the more flexible β-barrel structure if two additional β-strands are present in **NB4exp**. The quaternary structure in solution was also changed. The dimeric structure of **NB4** in the solid state (PDB code: 3wjb) was confirmed in solution by size exclusion chromatography. For **NB4exp**, the SEC experiment (performed by M.Sc. Alexander R. Grimm) revealed a monomeric structure in solution (Figure 25).

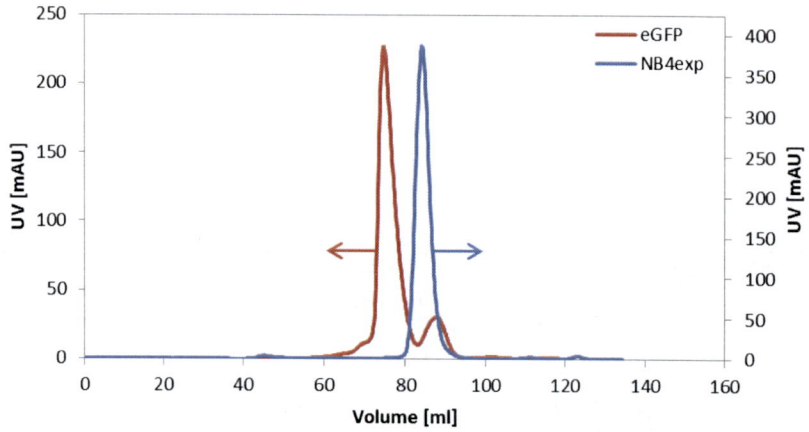

Figure 25. SEC trace of **NB4exp** and **eGFP** (M(**eGFP**) = 25 kDa; experiment performed by M.Sc. Alexander R. Grimm).

This monomeric structure should be less rigid than the dimeric assembly which explains its lower thermal stability.

B.2.3.2. Conjugation of the GH-type Catalysts to NB4exp

The new mutant was treated with the GH-type catalysts **C1-C3** (Scheme 16).

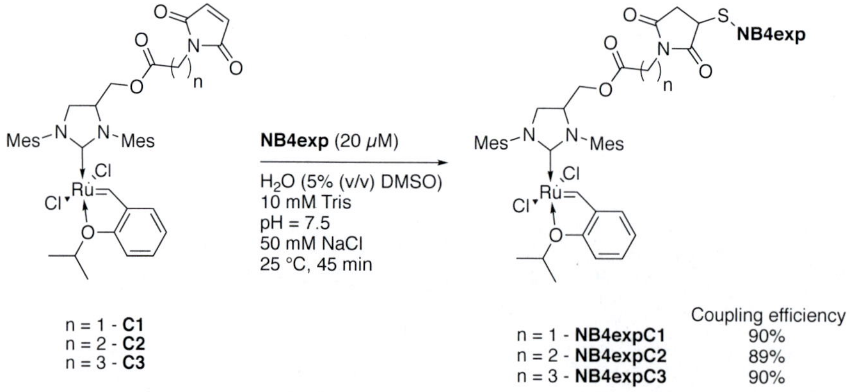

	Coupling efficiency
n = 1 - **NB4expC1**	90%
n = 2 - **NB4expC2**	89%
n = 3 - **NB4expC3**	90%

n = 1 - **C1**
n = 2 - **C2**
n = 3 - **C3**

Scheme 16. Coupling of GH-type catalysts to **NB4exp**.

With **NB4exp** as protein host, all catalysts undergo covalent anchoring in almost quantitative coupling efficiency. After purification, the biohybrid conjugates were analyzed by UV/vis spectroscopy (Figure 26).

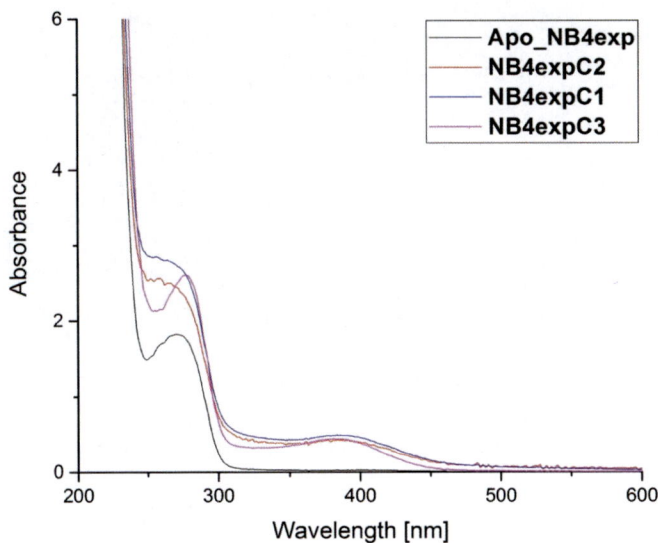

Figure 26. UV/vis spectra of **NB4exp** and **NB4exp** based biohybrid catalysts.

The UV/vis spectra show a maximum at 280 nm caused by the aromatic amino acids in the protein and a maximum at 380 nm due to MLCT from the ruthenium center to the NHC ligand.[16] Structural integrity was confirmed by CD spectroscopy. Each biohybrid catalyst based on **NB4exp** lead to the typical CD spectrum (Figure 27).

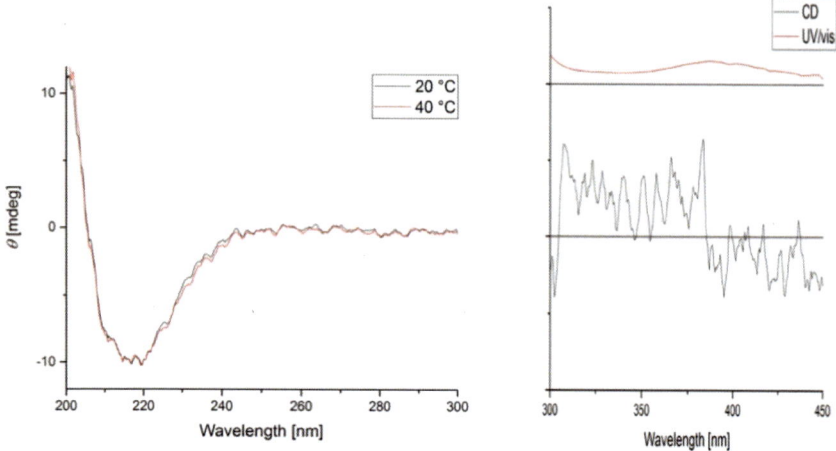

Figure 27. Left: CD spectra of **NB4exp_C1** at 20 °C and 40 °C; Right: CD spectrum of **NB4exp_C1** from 300-450 nm overlaid with the corresponding UV/vis spectrum.

Figure 27 shows a representative CD spectrum of **NB4exp_C1**. It corresponds to the typical β-barrel shape and indicates structural integrity upon conjugation. The molecular structure should also be stable under conditions used for catalysis (H_2O, pH = 6.0, 5 mM MES buffer, 200 mM NaCl, 40 °C). Unfortunately, no CD signal is induced at 380 nm. This is either caused by the racemic catalyst mixture used for conjugation, or shows a certain flexibility of the catalyst in the cavity of the protein.

ESI-TOF MS confirmed covalent anchoring. Values for the observed mass agree well with the calculated values (Table 13).

Table 13. ESI-TOF MS analysis for **NB4exp** and the biohybrid conjugates based on **NB4exp**.

Entry	Protein	m/z calculated	m/z found
1	NB4exp	22578.4254	22578.4254 ($[M^+]$)
2	NB4expC1	23371.5872	23319.6600 ($[M^+]$ -Cl_2H, +H_2O)
3	NB4expC2	23385.6028	23384.7882 ($[M^+]$ -H)
4	NB4expC3	23399.6185	23343.7970 ($[M^+]$ -Cl_2H, +H_2O)

In summary, the mutants of **NB** as well as **NB4exp** offer a stable second coordination sphere for the construction of artificial metatheases. The volume of the cavity is important to incorporate the catalyst. Table 14 summarizes the different in coupling efficiency between **NB4** and **NB4exp**.

Table 14. Coupling efficiency of nitrobindin based artificial metatheases.

Entry	Catalyst	Protein	Coupling efficiency[a] (%)
1	C1	NB4	0.
2	C1	NB4exp	90
3	C2	NB4	0
4	C2	NB4exp	89
5	C3	NB4	25
6	C3	NB4exp	90

[a] (c(metal)/c(protein)) • 100%.

B.2.4. Alkene Metathesis Reactions

B.2.4.1. Ring-opening Metathesis Polymerization

The biohybrid catalysts were investigated in the ROMP reaction of the 7-oxanorbornene **16**. The protein catalysts were compared to the commercially available and water-soluble metathesis catalyst **AquaMet**[12,18] and to the catalyst **COH** without linking unit (Table 15; for the chemical structure of both catalysts see B.2.6. Figure 33).

Table 15. Results of **NB** based artificial metatheases in the ROMP reaction of **16**.

Entry[a]	Catalyst	Protein	Conv.[b] [%]	TON	M_w^c x 10^3	PDI[c]	cis/trans[b]
1	AquaMet	-	16	1700	Oligomers	n.d.	46/54
2	COH	-	0	-	-	-	-
3[d]	COH	-	0	-	-	-	-
4[e]	COH	-	24	2400	n.d.	n.d.	70/30
5[f]	COH	-	41	4100	n.d.	n.d.	70/30
6[g]	COH	any NB	0	-	-	-	-
7[g]	AquaMet	any NB	0	-	-	-	-
8	C1	NB1	< 5	n.d.	n.d.	n.d.	n.d.
9	C2	NB1	19	2100	n.d.	n.d.	n.d.
10	C3	NB1	78	9700	200	1.05	43/57
11	C3	NB4	10	1100	Oligomers	n.d.	40/60
12	C1	NB4exp	25	3080	148, Mainly dimer and trimer	1.29	46/54
13	C2	NB4exp	81	10000	750	1.21	43/57
14	C3	NB4exp	75	9300	500	1.29	43/57

[a] c(**16**) = 0.2 M; catalyst loading is related to the metal content determinde by ICP-AES. [b] Determined by ^{1}H NMR spectroscopy. [c] Determined by GPC. [d] Catalysis in DMSO. [e] Catalysis in THF. [f] Catalysis in 2-propanol. [g] Catalyst is not specifically bound to the protein.

The ROMP reaction was carried out under slightly acidic pH values (pH = 6.0, 5 mM MES), since this reaction works best at decreased pH values. At pH = 7.4 (in 50 mM Tris buffer), no reaction occurs within 12 hours. The commercially available catalyst **AquaMet** gave 16% conversion with a catalyst loading of 0.008 mol % (Table 15, entry 1). Under similar conditions, the catalyst **COH** does not show any activity, maybe due to its insolubility in water (Table 15, entry 2). In the organic solvents DMSO, this catalyst was also inactive (Table 15, entry 3), but moderately active in THF and 2-propanol (Table 15, entries 4 and 5). If either one of the catalysts **COH** or **AquaMet** was used in the presence of an equimolar amount of protein (**NB1**, **NB4** or **NB4exp**), conversion was not observed at all (Table 15, entries 6 and 7). Due to the high number of functionalities on the protein surface, the catalyst is most likely bound to amino acid residues and therefore deactivated. Conversion was not observed, if the free catalyst was removed by HiTrap desalting column. **NB1_C1** showed a very low activity with a conversion less than 5% (Table 15, entry 8). The activity increased with longer spacer units. **NB1_C2** showed 19% conversion and **NB1_C3** showed 78% conversion (Table 15, entries 9 and 10). This was explained by the accessibility of the active site. The catalysts with shorter spacer should be buried in the cavity and be less flexible. The active center should be further shielded by the growing polymer chain, which makes it difficult for a new monomer to approach the active site. This explains the only low activity of **NB4** with catalyst **C3**. Only 10% conversion was observed (Table 15, entry 11). With the increased cavity of **NB4exp**, all three catalysts showed a higher activity than **NB4**. Both catalysts with longer spacers gave similar conversions between 75 and 81% (Table 15, entries 12 and 13). The highest molecular weight was observed for **NB4exp_C2**. The larger cavity volume seems to benefit polymer growth in the aqueous media. With **NB4exp_C1**, the conversion dropped to 25% (Table 15, entry 14). Beside a polymeric fraction (M_w = 148.000 g/mol), oligomers were also observed. A shorter linker restricted the flexibility within the cavity and moved the metal center from the loop region to the β-sheets, where the influence of the protein is increased. This may lead to a lower catalytic activity and therefore lower conversion; interactions of certain amino acid residues may also lead to chain transfer or chain termination *via* backbiting of the catalyst (*cf.* B.2.6. Figure 34). GPC analysis showed the formation of oligomers. A crystal structure would be needed to show the closest residues and their influence on the active site. Nevertheless, the protein sphere drastically influences the activity at very low catalyst loadings. The protein makes the active species fully water-soluble and provides shielding against strong and undefined coordinative residues/molecules. During the reaction, the polymer chain precipitates from the aqueous phase. Here, the protein may assist with the phase transfer by either keeping the active site in solution or by accumulating further substrates within the cavity even after precipitation.

Kinetic investigations were performed with **NB4exp_C2** and **AquaMet** as water-soluble protein free variant (Figure 28 and Figure 29).

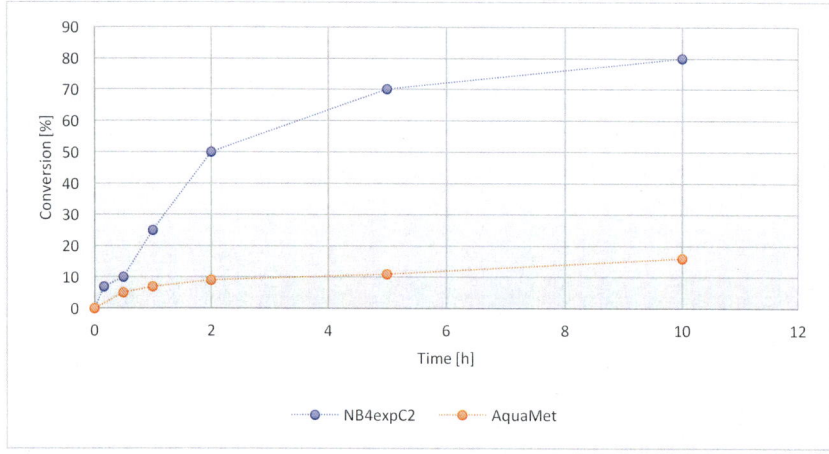

Figure 28. Time-conversion plot of **NB4exp_C2** and **AquaMet** in the ROMP of **16**.

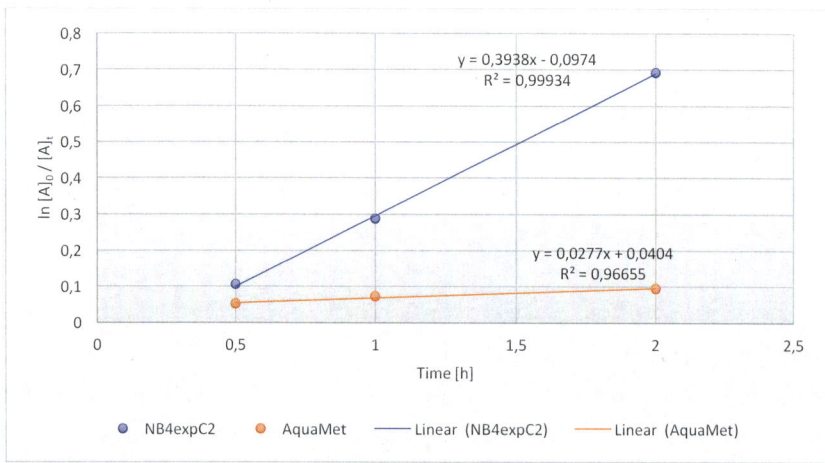

Figure 29. Semi-logarithmic plot of the concentration of **16** vs time.

Semi-logarithmic representation of the time-conversion leads to a linear plot (Figure 29). The obtained values k_{obs} of **NB4exp_C2** and **AquaMet** show that the biohybrid catalyst converts substrate **16** ca. 10 times faster (k_{obs}(**NB4exp_C2**) = 0.40 h^{-1}; k_{obs}(**AquaMet**) = 0.03 h^{-1}). The selectivity over time stays constant for all catalysts shown in Table 15.

The reactivity of the charged norbornene **26** was also investigated (Table 16).

Table 16. ROMP of **26**.

Entry[a]	Catalyst	Protein	Conv.[b] [%]	TON
1	AquaMet	-	99	12300
2	AquaMet	Any NB	0	-
3	COH	Any NB	0	-
4	C3	NB1	0	-
5	C3	NB4	0	-
6	C3	NB4exp	70	8700
7	C2	NB4exp	72	8900

[a] $c(\mathbf{26})$ = 0.2 M; catalyst loading is related to the metal content determinde by ICP-AES. [b] Determined by ^{1}H NMR spectroscopy.

In organic solvents, charged substrates are not polymerized with high conversion by metal catalysis because the highly charged polymers would be insoluble. But such polymers would be of high interest due to their potential use as polyelectrolytes. The protein catalysts based on **NB4exp** allow high conversion of norbornene **26** (Table 16, entries 7 and 8), whereas the protein catalysts based on **NB1** and **NB4** do not show any conversion (Table 16, entries 5 and 6). This may be due to electronic repulsion with a charged lysine residue K127. In the larger cavity of **NB4exp**, the distance between lysine and ruthenium is longer and therefore enables the ROMP of substrate **26** (Figure 30).

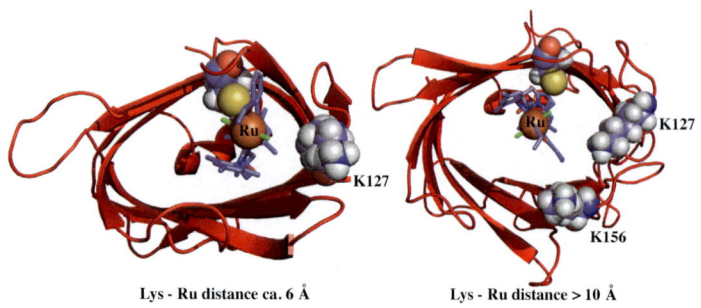

Figure 30. Ru-Lys distance in **NB4_C3** (left) and **NB4exp_C3** (right) calculated with YASARA.

B.2.4.2. Ring-closing Metathesis

The RCM reaction was utilized before as benchmark reaction for the first artificial metatheases.[1,5,10] The substrate diallyl tosyl amine **3** was chosen, although it is completely insoluble in water.[19-20] Table 17 shows the results of the RCM reaction of **3** with nitrobindin immobilized GH-type catalysts.

Table 17. RCM of diallyl tosyl amine **3**.

Tos–N(allyl)₂ **3** → catalyst (1.0 mol %), H₂O (5 mM MES, 200 mM NaCl), pH = 6.0, 25 °C, 750 mbar, 12 h, $-C_2H_4$ → Tos-pyrroline **4**

Entry[a]	Catalyst	Protein	Conv.[b] [%]	TON
1	AquaMet	-	41	41
2	AquaMet	Any NB	0	-
3	COH	Any NB	0	-
4	C3	NB1	0	-
5	C3	NB4	0	-
6	C3	NB4exp	35	35
7	C2	NB4exp	35	35
8	C1	NB4exp	16	16

[a] c(**3**) = 0.05 M; catalyst loading is related to the metal content determinde by ICP-AES. [b] Determined by ¹H NMR spectroscopy.

AquaMet catalyzed the RCM reaction of **3** with 41% conversion (Table 17, entry 1), like other catalysts under mild conditions.[9] Conversion was not observed if the metal catalyst **AquaMet** or **COH** was used together with a protein but not being conjugated (Table 17, entries

2-3). In this case, the catalyst should be deactivated through coordination of the protein surface.[5] If catalysts were used that are based on **NB1** and **NB4**, the protein precipitated after the substrate was added and the RCM product was not formed (Table 17, entries 4-5). The protein even precipitated if organic co-solvents (DMSO and THF up to 30% (v/v)) were used. This may be caused by the properties of the quaternary structure where the dimeric structure of **NB4** or **NB1** is mainly stabilized by hydrophobic amino acid residues (Figure 31).

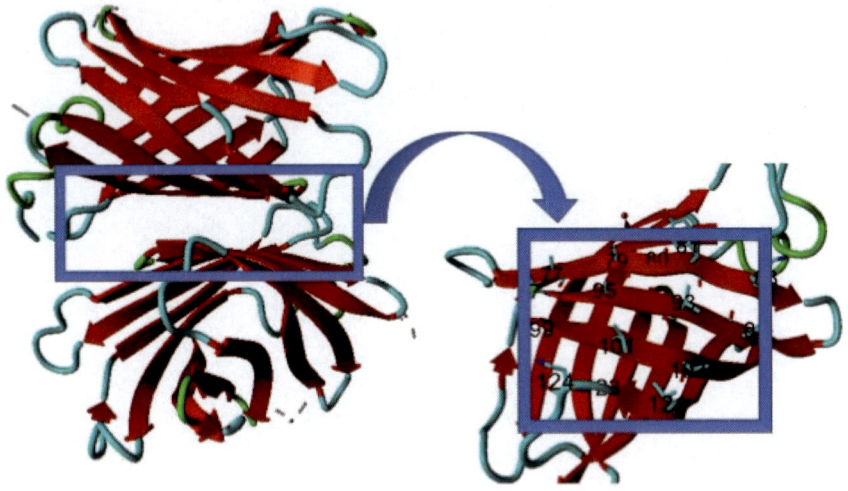

Figure 31. Hydrophobic dimeric interface of NB4. The main contact within the interface of the two NB units is made by A77, G80, V93, A95, G99, V191, V103, L121 and G123.

The hydrophobic substrate might interact with this part of the protein and cleave the dimeric structure, making the whole protein structure less stable and lead to precipitation. The monomeric **NB4exp** seems to contain a suitable cavity for such nonpolar substrates, since precipitation was not observed. The biohybrid conjugates with **NB4exp** can give TON up to 35 (Table 17, entries 6-8). The present system is nearly twice as active as the artificial metatheases described in chapters A and B1 (Ward and coworkers[11]: TON (hCAII) = 25, TON (Avi) = 20; Hilvert and coworkers[7]: TON (MjHSP) = 25; Klein Gebbink and coworkers[14]: TON (lipase) = 20; Schwaneberg, Okuda and coworkers[8]: TON (**FhuA**) = 10; Matsuo et al.[16]: TON (α-chymotrypsin) = 20). But the results cannot be directly compared because the buffering conditions and co-solvents vary.

Both protein hosts **NB1/NB4** and **NB4exp** show high activities with a substrate that is fully soluble in water (Table 18).

Table 18. RCM of diol 18.

HO–◇–OH (18) → catalyst (1.0 mol %), H₂O (5 mM MES, 200 mM NaCl), pH = 6.0, 25 °C, 750 mbar, 12 h, -C₂H₄ → HO–△–OH (19)

Entry[a]	Catalyst	Protein	Conv.[b] [%]	TON
1	AquaMet	-	> 99	100
2	AquaMet	Any NB	0	-
3	COH	Any NB	0	-
4	C3	NB1	> 99	100
5	C2	NB1	89	89
6	C3	NB4	69	69
7	C3	NB4exp	> 99	35
8	C2	NB4exp	> 99	35
9	C1	NB4exp	45	45

[a] c(3) = 0.05 M; catalyst loading is related to the metal content determinde by ICP-AES. [b] Determined by ¹H NMR spectroscopy.

In contrast to the insoluble amine **3**, the diol **18** that is completely soluble in water did not cause precipitation. The protein free catalysts **AquaMet** allowed full conversion within 12 hours (Table 18, entry 1). Where the catalyst was not specifically bound within the cavity, conversion was not observed at all (Table 18, entries 2 and 3). The proteins with longer spacer and larger cavity also led to full conversion within 12 hours (Table 18, entries 4, 7-8). The combination of the catalysts **NB1** and **C2** gave a slightly higher conversion than **NB4** and **C3** (TON = 89 vs. TON = 69, Table 18, entries 5-6). This effect was stronger in **NB4exp** (TON = 100), but the catalyst **C1** with the short spacer only showed moderate activity of TON = 45 (Table 18, entry 9). Why the TON is so low cannot be explained at this stage, but it may be caused by restricted access of the substrate or by a small degree of freedom for the catalyst with short spacer.

The reaction of an umbelliferone precursor **28** to umbelliferone **29** *via* RCM was investigated (Scheme 17). Since the product of this reaction shows a fluorescence, the reaction was monitored with a Tecan 96-well plate fluorescence reader (Figure 32).

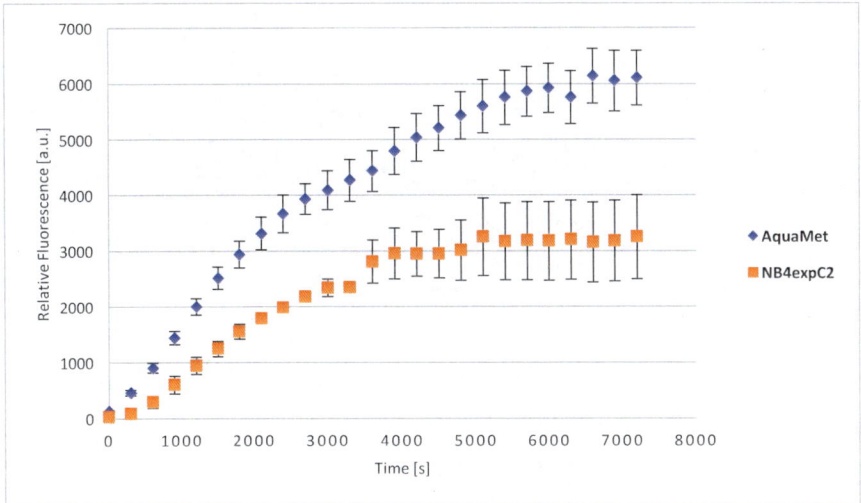

Scheme 17. RCM of **28** to umbelliferone **29**.

Figure 32. Fluorescence measurement during the RCM of **28** (c(**28**) = 10 mM).

Figure 32 shows the results of the RCM reaction of **28** to the fluorescent umbelliferone Because the product is insoluble in water, the fluorescence signal could not be calibrated. Nevertheless, the results can be compared qualitatively and the protocol can be applied for the screening of mutants of the **NB4exp**, as shown for Sav/Avi by the Ward and coworkers.[12] The catalyst **AquaMet** is converting the substrate in a constant matter, following a first order kinetics in the first instance. With the protein catalyst **NB4expC2** however, an induction period is observed. After two hours, approximately half as much substrate is converted than with **AquaMet**. The lower conversion is ascribed to a product inhibition of the protein catalyst. The hydrophobic product formed during the conversion is expected to stay in the cavity where it is blocking the space for new substrate. Figure 35 shows the interaction of the product with the protein. A full list of blank runs is included in the experimental section. Without any GH-type catalyst, the fluorescence is not enhanced (*cf.* Figure 36). These observations were expected, because the RCM reaction has to be catalyzed by a metal center.

B.2.4.3. Cross Metathesis

Cross metathesis (CM) of terminal olefins was targeted with the water-soluble allyl alcohol 30, the water insoluble styrene 31 and the partially water-soluble *para*-methoxy styrene 32 (Table 19).

Table 19. CM reaction catalyzed by nitrobindin based biohybrid catalysts.

$$2\ R\diagup\!\!\!\!\diagdown \xrightarrow[\substack{H_2O\ (5\ mM\ MES,\ 200\ mM\ NaCl)\\ pH = 6.0\\ 40\ °C,\ 750\ mbar,\ 24\ h,\ -C_2H_4}]{\text{catalyst (1 mol \%)}} R\diagup\!\!\!\!\diagdown\!\!\!\!\diagup\!\!\!\!\diagdown R$$

R = CH$_2$OH - 30
R = Ph - 31
R = C$_6$H$_4$-*p*-OMe - 32

R = CH$_2$OH - 33
R = Ph - 34
R = C$_6$H$_4$-*p*-OMe - 35

Entry[a]	Catalyst	Protein	Substrate	Conv.[b] [%]	E/Z[b]	TON
1	AquaMet	Any NB	30, 31, 32	0	-	0
2	COH	Any NB	30, 31, 32	0	-	0
3	AquaMet	-	30	79	20/1	79
4	C3	NB1	30	> 99	20/1	100
5	C3	NB4	30	69	20/1	69
6	C3	NB4exp	30	> 99	20/1	100
7	C2	NB4exp	30	> 99	20/1	100
8	C1	NB4exp	30	45	20/1	45
9	AquaMet	-	31	98	99/1	98
10	C3	NB1	31	> 99	99/1	100
11	C3	NB4	31	50	99/1	50
12	C3	NB4exp	31	> 99	99/1	100
13	C2	NB4exp	31	> 99	99/1	100
14	C1	NB4exp	31	45	99/1	45
15	AquaMet	-	32	94	99/1	94
16	C3	NB1	32	> 99	99/1	100
17	C3	NB4	32	45	99/1	45
18	C3	NB4exp	32	> 99	99/1	100
19	C2	NB4exp	32	> 99	99/1	100
20	C1	NB4exp	32	40	99/1	40

[a] c(substrate) = 0.05 M; catalyst loading is related to the metal content determinde by ICP-AES. [b] Determined by ^{1}H NMR spectroscopy.

All three substrates undergo CM reaction in aqueous buffer solution. For the styrene derivatives that are not (**31**) or just partially (**32**) water-soluble, the conditions were rather emulsion like. Conversion was not observed if the protein free catalysts **COH** or **AquaMet** were used in a solution with an equimolar amount of protein (Table 19, entries 1-2). This agrees with the results in ROMP and RCM. **AquaMet** moderately converted allyl alcohol **30** (Table 19, entry 3).[18] The protein based catalysts showed moderate to excellent conversion. A trend was observed in terms of cavity space and linker length. With **NB4exp**, the activity decreased with the length of the spacer (Table 19, entries 6-8). The activity was also lower with **NB4** where the cavity is smaller (Table 19, entry 5). The selectivity was nearly the same as with **AquaMet**. The same trends were observed for the other substrates used in this reaction (substrate **31**, Table 19, entries 9-14; substrate **32**, Table 19, entries 15-20).

B.2.5. Summary and Conclusion

In this chapter, the use of the small β-barrel protein nitrobindin as scaffold for artificial metatheases is shown. **NB1** and **NB4** mutant showed higher conversions in the RCM of amine **3** than other artificial metatheases. The volume of the cavity turned out crucial for the incorporation of sterically demanding catalysts. Best results were obtained with the nitrobindin mutant **NB4exp** that was provided by the research group of Prof. Schwaneberg and which contains an enlarged cavity due to two additional β-strands. All investigated catalysts could be incorporated into this protein with high coupling efficiency, even **C1** with the shortest linker. The absence of an additional CD signal indicates that the catalyst seems to be relatively flexible in the cavity environment.

All catalysts are more or at least as active as the water-soluble and commercially available catalyst **AquaMet**. The biohybrid catalysts showed good to excellent activities in ROMP, RCM and CM reactions. This is the only artificial metathease system which actively catalyzes all of these reactions. The umbelliferone assay will be useful to screen mutants of **NB4exp** to enhance the activity even more.

B.2.6. Experimental Section

General Comments

The proteins **NB1** and **NB4** were provided by the group of Prof. Hayashi, Osaka University. The **NB4exp** variant was provided by the group of Prof. Schwaneberg, RWTH Aachen University.

All experiments were performed under a nitrogen or argon atmosphere using standard Schlenk techniques or an MBraun nitrogen-filled glovebox. DMSO was degassed by using "pump- freeze-thaw" cycles. ^{1}H NMR spectra were recorded on a Bruker Avance III 400 NMR spectrometer. Chemical shifts are reported in ppm relative to the residual solvent resonances. MALDI-TOF MS analyses were performed on a Bruker autoflex III mass spectrometer. ESI-TOF MS analysis was performed on a Bruker microTOF focus III spectrometer. UV/vis spectra were recorded on a Shimadzu BioSpec-nano spectrometer. Circular dichroism (CD) spectra were recorded on a JASCO J720S spectrometer. VT_CD spectra were recorded on a *JASCO* J-1100 spectrometer equipped with a single position Peltier cell holder. The pH values were measured with a Horiba F52 pH meter. Gel permeation chromatography (GPC) was performed on a TOSOH SC8020 apparatus with a refractive index (RI) detector with a TOSOH TSKGel G5000H HR column. Gas chromatography (GC) was performed on a Shimadzu GC-2014. Size Exclusion Chromatography (SEC) was performed with GE Healthcare Sephadex G-25 medium. Inductively Coupled Plasma Atomic Emission Spectroscopy (ICP-AES) measurements were performed on a Shimadzu ICPS-7510 spectrometer. A serial dilution was prepared by using a commercially available ruthenium standard (0.992 µg/mL; Fluka Analytical, Sigma-Aldrich catalog number: 207446-100ML). The calibration was done with samples containing 0.0 µM, 0.5 µM, 1 µM and 2 µM Ru. The protein sample was diluted to 1.5 µM protein concentrations. Each batch biohybrid conjugate was analyzed prior to use in catalysis. Fluorescence measurements were performed on a Tecan 96-well plate reader. The temperature was adjusted to 40 °C. The excitation wavelength was λ_{ex} = 322 nm. The emission wavelength was λ_{em} = 440 nm. Grubbs-Hoveyda type complexes **COH**, **C1**, **C2** and **C3**, Substrates **3, 16, 18, 26** and **28** were synthesized as previously reported. **AquaMet**, Substrates **30, 31, 32** and umbelliferone **29** are commercially available and were used as received.

General procedure for the conjugation of the Grubbs-Hoveyda type complexes.

To a solution containing the freshly reduced (with DTT) nitrobindin variant (10 µM) in Tris buffer (10 µM, pH 7.5, 50 µM NaCl), complex **C3** (10 equiv.) in DMSO (1% (v/v)) was added. Complex **C1** or **C2** (10 equiv. respectively) were added in 10% (v/v) DMSO. After 30 min, the reaction mixture was centrifuged to remove precipitates, the supernatant was concentrated (ca. 1 mL) by using an amicon ultrafiltration cell (10 kDa MWCO). If more than 1% (v/v) DMSO was used, the solution was diluted to 5 mL and concentrated again (2 x). The proteins were purified with a sephadex column (GE Healthcare) with Tris buffer (pH 7.5, 10 mM, 200 mM NaCl) as an eluent.

Trace metal analysis of the washing solution of a blank run with protein (10 µM, pH 7.5, 50 µM NaCl) and **COH** (10 equiv.) in DMSO (1% (v/v)) did not show detectable Ru after purification.

General procedure for ROMP of oxanorbornene derivative 16 and norbornene derivative 26.

The freshly prepared and filtered biohybrid catalyst (500 µL of a 16 µM solution, based on the ruthenium content as determined by ICP-AES) was subjected to a HiTrap desalting column to exchange the buffer to either Tris buffer solution (pH 7.5, 10 mM, 200 mM NaCl) or in MES buffer solution (5 mM, pH 6.0, 200 mM NaCl). The norbornene derivative **16** was added via syringe (15 µL, 0.2 M) to the solution containing the biohybrid catalysts. The norbornene derivative **26** was dissolved in a low amount of buffer solution and added via syringe (final concentration: 0.2 M). After reaction time of 12 h or 24 h, 0.5 mL of THF and 0.5 mL of ethylvinylether were added and the solution was mixed on a vortex and incubated at room temperature for 20 min. The substrate **16** and polymer **17** were extracted with chloroform, the organic phase was filtered, dried over $MgSO_4$, and analyzed by ^1H NMR spectroscopy in $CDCl_3$. The conversion and the *cis/trans* ratio was determined by the method reported by Feast and Harrision.[19] The molecular weight was determined *via* GPC. For substrate **26**, the solvent was removed *in vacuo* and the residue was taken up in D_2O and analyzed like reported.[20-21]

General procedure for RCM.

The freshly prepared and filtered biohybrid catalyst (1 mL of a 1.25 mM solution, based on the ruthenium content as determined by ICP-AES) was subjected to a HiTrap desalting column to exchange the buffer to MES buffer solution (pH 6.0, 5 mM, 200 mM NaCl). Amine **3** or diol **18** were added via syringe (125 mmol) and the reaction mixture was heated to 40 °C. After reaction time of 24 h 0.5 mL THF and 0.5 mL ethylvinylether were added and the solution was mixed on a vortex and incubated at room temperature for 20 min. The substrate and the product were extracted with chloroform, the organic phase was filtered, dried over $MgSO_4$, and analyzed by ^1H NMR spectroscopy and GC.[22]

General procedure for CM.

The freshly prepared and filtered biohybrid catalyst (1 mL of a 1.25 mM solution, based on the ruthenium content as determined by ICP-AES) was subjected to a HiTrap desalting column to exchange the buffer to MES buffer solution (pH 6.0, 5 mM, 200 mM NaCl). Substrates **30**, **31** or **32** were added via syringe (125 mmol) and the reaction mixture was heated to 40 °C. After reaction time of 24 h 0.5 mL THF and 0.5 mL ethylvinylether were added and the solution was mixed on a vortex and incubated at room temperature for 20 min. The substrate and the product were extracted with chloroform, the organic phase was filtered, dried over $MgSO_4$, and analyzed by ^1H NMR spectroscopy and GC.

Molecular Modeling

Modeling of the apo-NB variants, **NB4**, **NB11**, was performed as reported previously[2-3] using YASARA[23] Structure Vers. 13.6.16, employing force field AMBER03.[24] As structural basis for the modeling of the biohybrid catalysts, the X-ray structure of native nitrobindin (PDB code: 2A13), and the NB variants (**NB4** (PDB code: 3WJB) and **NB1** (PDB: 4YMY)) were used. The model for NB4exp was kindly provided by Dr. Marco Bocola (group of Prof. Schwaneberg, Institute for Biotechnology, RWTH Aachen University). According to the previously published procedure, the modeling was carried out using YASARA Structure Vers. 13.6.16 employing force field AMBER03 for protein residues and GAFF[25] using AM1/BCC[26] partial charges for the catalyst covalently bound to Cys96. The metal was replaced by cobalt, since no parameters are available for ruthenium. To maintain the correct coordination geometry, the distances and angles from the metal to all coordinating atoms were constrained according to the X-ray structure of the Grubbs-Hoveyda[27] catalyst by force field arrows. The charge of the metal was set to +2 and the total charge of the catalyst was set to zero. The linker was placed manually in the cavity adjacent to Cys96 and a bond from Cys S atom to the C1 atom of the maleimide group was defined, according to the linker geometry in the high-resolution X-ray structure of the ruthenium complex. The constructed hybrid catalysts were solvated in a box of TIP3P water molecules using periodic boundaries at pH 7 and a density of 0.997 g/mL. Three starting structures were analyzed and favorable models were identified for covalent attachment to the reactive maleimide atoms by steepest descent minimization and simulated annealing. The pre- minimized structures were relaxed using molecular dynamics calculations at 298 K for 5000 ps and snapshots were taken every 25 ps to analyze the binding modes. Van der Waals volume of the ligand was calculated using YASARA.

Chemical structure of COH and AquaMet

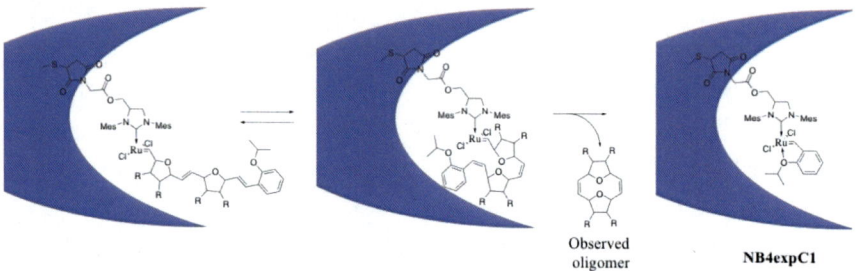

COH **AquaMet**

Figure 33. COH and AquaMet.

ROMP of 16 – Oligomer Formation

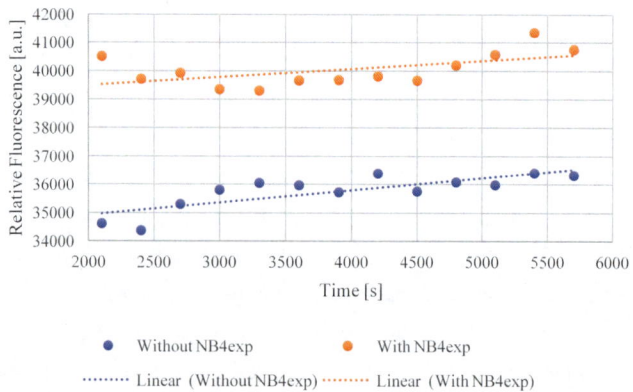

Figure 34. Possible formation of oligomers *via* backbiting.

RCM of 28 – Fluorescence Experiments

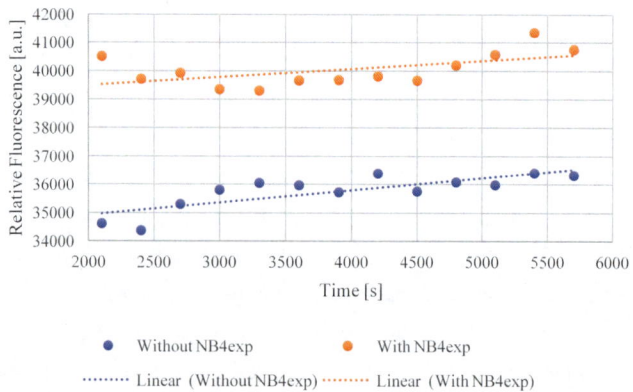

Figure 35. Fluorescence measurement of **29** in the present (orange) and the absence (blue) of **NB4exp**.

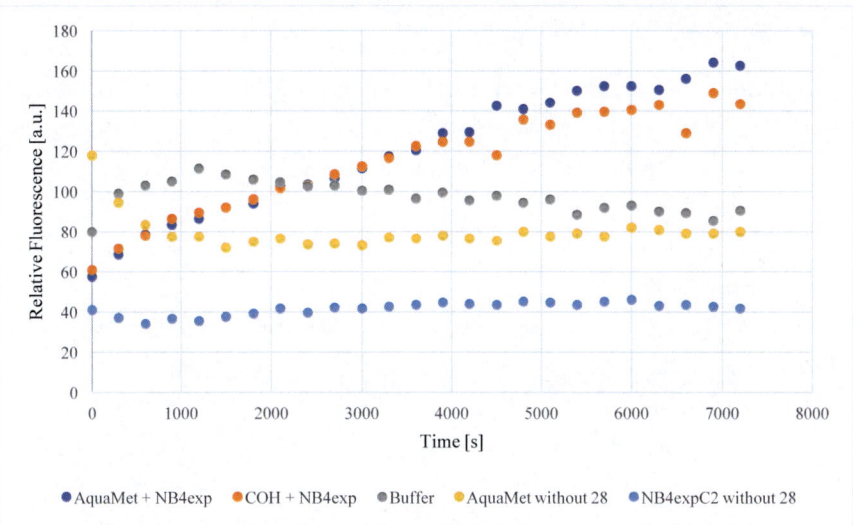

Figure 36. Blank experiments of substrate **28** in different media and catalysts without substrate.

B.2.7. References

(1) Bianchetti, C. M.; Blouin, G. C.; Bitto, E.; Olson, J. S.; Phillips, G. N. *Proteins: Struct., Funct., Bioinf.* **2010**, *78*, 917.

(2) Onoda, A.; Fukumoto, K.; Arlt, M.; Bocola, M.; Schwaneberg, U.; Hayashi, T. *Chem. Commun.* **2012**, *48*, 9756.

(3) Fukumoto, K.; Onoda, A.; Mizohata, E.; Bocola, M.; Inoue, T.; Schwaneberg, U.; Hayashi, T. *ChemCatChem* **2014**, *6*, 1229.

(4) Onoda, A.; Kihara, Y.; Fukumoto, K.; Sano, Y.; Hayashi, T. *ACS Catal.* **2014**, *4*, 2645.

(5) Himiyama, T.; Sauer, D. F.; Onoda, A.; Spaniol, T. P.; Okuda, J.; Hayashi, T. *J. Inorg. Biochem.* **2016**, *158*, 55.

(6) Zhang, C.; Srivastava, P.; Ellis-Guardiola, K.; Lewis, J. C. *Tetrahedron* **2014**, *70*, 4245.

(7) Mayer, C.; Gillingham, D. G.; Ward, T. R.; Hilvert, D. *Chem. Commun.* **2011**, *47*, 12068.

(8) Philippart, F.; Arlt, M.; Gotzen, S.; Tenne, S.-J.; Bocola, M.; Chen, H.-H.; Zhu, L.; Schwaneberg, U.; Okuda, J. *Chem. Eur. J.* **2013**, *19*, 13865.

(9) Sauer, D. F.; Bocola, M.; Broglia, C.; Arlt, M.; Zhu, L.-L.; Brocker, M.; Schwaneberg, U.; Okuda, J. *Chem. Asian J.* **2015**, *10*, 177.

(10) Zhu, L.; Arlt, M.; Liu, H.; Bocola, M.; Sauer, D. F.; Gotzen, S.; Okuda, J.; Schwaneberg, U. In *Bio-Synthetic Hybrid Materials and Bionanoparticles: A Biological Chemical Approach Towards Material Science*; The Royal Society of Chemistry: **2015**, p 57.

(11) Lo, C.; Ringenberg, M. R.; Gnandt, D.; Wilson, Y.; Ward, T. R. *Chem. Commun.* **2011**, *47*, 12065.

(12) Jeschek, M.; Reuter, R.; Heinisch, T.; Trindler, C.; Klehr, J.; Panke, S.; Ward, T. R. *Nature* **2016**, *537*, 661.

(13) Mallin, H.; Hestericova, M.; Reuter, R.; Ward, T. R. *Nat. Protocols* **2016**, *11*, 835.

(14) Basauri-Molina, M.; Verhoeven, D. G. A.; van Schaik, A. J.; Kleijn, H.; Klein Gebbink, R. J. M. *Chem. Eur. J.* **2015**, *21*, 15676.

(15) Zhao, J.; Kajetanowicz, A.; Ward, T. R. *Org. Biomol. Chem.* **2015**, *13*, 5652.

(16) Matsuo, T.; Imai, C.; Yoshida, T.; Saito, T.; Hayashi, T.; Hirota, S. *Chem. Commun.* **2012**, *48*, 1662.

(17) Wallace, B. A.; Lees, J. G.; Orry, A. J. W.; Lobley, A.; Janes, R. W. *Protein Sci.* **2003**, *12*, 875.

(18) Skowerski, K.; Szczepaniak, G.; Wierzbicka, C.; Gulajski, L.; Bieniek, M.; Grela, K. *Catal. Sci. Tech.* **2012**, *2*, 2424.

(19) Feast, W. J.; Harrison, D. B. *Polym. Bull.* **1991**, *25*, 343.
(20) Jordan, J. P.; Grubbs, R. H. *Angew. Chem.* **2007**, *119*, 5244.
(21) Jordan, J. P.; Grubbs, R. H. *Angew. Chem. Int. Ed.* **2007**, *46*, 5152.
(22) Gotzen, S. Doctoral Dissertation, RWTH Aachen University, **2016**.
(23) Krieger, E.; Darden, T.; Nabuurs, S. B.; Finkelstein, A.; Vriend, G. *Proteins: Struct., Funct., Bioinf.* **2004**, *57*, 678.
(24) Duan, Y.; Wu, C.; Chowdhury, S.; Lee, M. C.; Xiong, G.; Zhang, W.; Yang, R.; Cieplak, P.; Luo, R.; Lee, T.; Caldwell, J.; Wang, J.; Kollman, P. *J. Comput. Chem.* **2003**, *24*, 1999.
(25) Wang, J.; Wolf, R. M.; Caldwell, J. W.; Kollman, P. A.; Case, D. A. *J. Comput. Chem.* **2004**, *25*, 1157.
(26) Jakalian, A.; Jack, D. B.; Bayly, C. I. *J. Comput. Chem.* **2002**, *23*, 1623.
(27) Garber, S. B.; Kingsbury, J. S.; Gray, B. L.; Hoveyda, A. H. *J. Am. Chem. Soc.* **2000**, *122*, 8168.

B.3. Whole-Cell Catalysis Based on β-Barrel Proteins FhuA and Nitrobindin

B.3.1. Introduction

The solubilized membrane protein **FhuA** and/or the water-soluble proteins **NB** or **NB4exp** (Chapters B.1. and B.2.) can be used for biohybrid catalysis after they are purified. In this work, the protein purification is only mentioned briefly and purification of especially membrane proteins such as FhuA is costly and time consuming.[1-3] Therefore, designing methods and strategies to avoid time consuming purifications are desirable to for a cost-effective production of chemicals. Nearly all industrial processes for bulk chemical production by biotechnological means employ whole-cell systems for conversions due to cost-benefits. Such industrially applied whole-cell catalysts carry the overexpressed target protein or enzyme that catalyzes the desired reaction, e.g. transfer hydrogenation.[4-7] This strategy does not only save time and costs for the purification of the protein. Extraordinary selectivity has been obtained and the catalyst was easily removed by filtration or sedimentation of the cells. Regeneration of the co-factor (e.g. NADH or NADPH) that is usually required by the enzyme can be achieved by introducing a co-factor regeneration. Coupled regeneration to the organism metabolism ensures highly efficient co-factor regeneration. Whole-cells protect the target enzyme against denaturation and external influences. Reactions may occur even in pure organic solvent or in the neat substrate. Organic solvents expand the scope to substrates that are insoluble or instable in water. A biocatalytic system that allows to regenerate the co-factor and works with neat substrate is found in the asymmetric reduction of ketones, where a NADH-dependent carbonyl reductase from *Candida parapsilosis* (CPCR) was overexpressed in *E. coli* as host cell.[6] The cofactor was regenerated by isopropanol-coupled co-factor regeneration. Only one enantiomer was found as expected for a biocatalyst. This approach gave higher conversion than other systems (Scheme 18).[6]

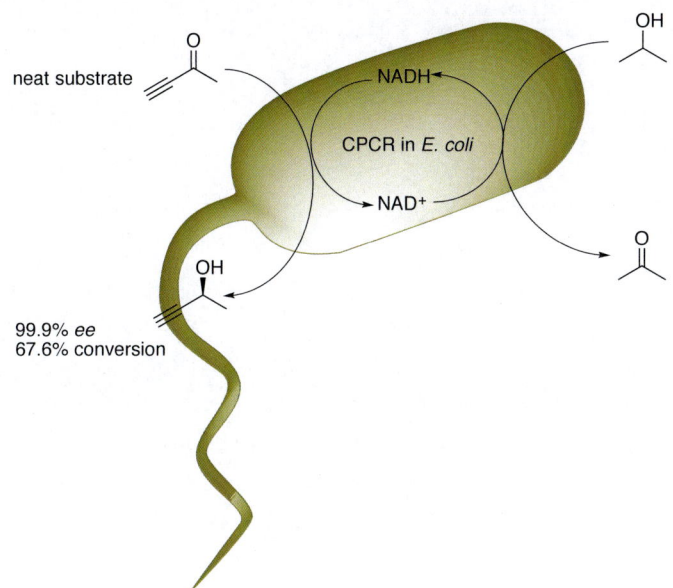

Scheme 18. Whole-cell biocatalytic system based on CPCR expressed in *E. coli*.[6] *E. coli* cells were lyophilized and partially rehydrated to achieve high productivity.[6]

Artificial whole-cell catalysis should lead to new tailor-made catalysts by combining the advantage of biocatalysis with improved practical implementation. This chapter describes two approaches to artificial whole-cell catalysts. Both differ fundamentally from the system presented by Ward and coworkers (*cf.* Chapter A).[8]

An artificial whole-cell catalyst for metathesis reactions is to be created with the transmembrane protein FhuA. Since this is located within the outer membrane of *E. coli*, it should be easily accessible for the catalyst to conjugate and for the substrate to react. The activity and the selectivity is to be compared to the isolated protein systems presented in chapter B.1.

As a second approach, the soluble protein nitrobindin is to be immobilized on the surface of *E. coli* using a cell-surface display (CSD) technology. In cooperation with the research group of Prof. Schwaneberg at RWTH Aachen University, M.Sc. Alexander R. Grimm carries out the biotechnological experiments. Figures were jointly prepared by M.Sc. Alexander R. Grimm and Prof. Schwaneberg. Goal is a rhodium-based NB CSD catalyst capable of catalyzing the polymerization of phenylacetylene. This is to be compared to the soluble system of Hayashi and coworkers (*cf.* B.2.1.).[9-10]

B.3.2. Whole-Cell Metathesis Catalysts Based on the Transmembrane Protein FhuA

Instead of extracting the **FhuA** from *E. coli*,[1,11-12] the whole-cell has been used to anchor the catalyst. The cells were received as lyophilized powder containing the **FhuA_ΔCVF**[tev] variant in the outer membrane. First approaches to anchor the catalysts directly to the lyophilized cells without swelling or adjusting the water concentration by vapor diffusion[6] failed. A re-hydration strategy was tried as reported before.[6] Therefore, different water concentrations were adjusted by using a vapor diffusion method. The cells were exposed to a water atmosphere generated by saturated water with different salts (KCl, NaCl and KNO$_3$). This approach led to catalytic results in the ROMP reaction of oxanorbornene **16**, but the results were not reproducible. The cells were placed directly into a buffer solution for two to three days. An aqueous buffer solution consisting of 100 mM NaCl at pH = 8 was used to avoid osmotic shock for the cell. The coupling procedure used for the solubilized **FhuA** was applied to the whole-cell system (Scheme 19).

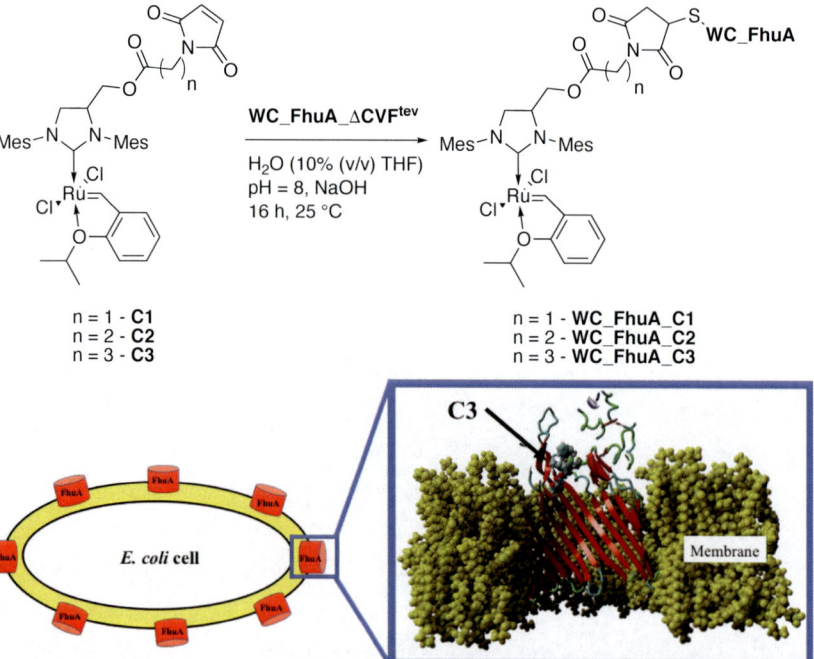

Scheme 19. Top: Coupling of **C1-C3** to **WC_FhuA**. Bottom: Illustration of the **WC_FhuA_C3** in the *E. coli*.

After incubation, the cell suspension was centrifuged. The supernatant was discarded and a mixture of buffer and THF 1:1 was supplemented, the cell pellet was suspended and centrifuged again. This procedure was repeated three times to remove the unbound GH type catalyst. Afterwards, the cells where washed with buffer only by using the same method to remove the THF. Finally, the cells were washed two times with the reaction buffer. The whole-cells were then ready for being employed in whole-cell catalysis (*cf.* chapter B1).

The generated whole-cell catalysts were applied in the ROMP reaction of the benchmark 7-oxanorbornene substrate **16**. The whole-cell catalysts with **FhuA_ΔCVF^{tev}** and a deletion variant not containing C545 were used (Table 20).

Table 20. ROMP results with whole-cell metathease.

Entry^a	Catalyst	Conv.^{b} [%]	cis/trans^b	TON
1	WC_FhuA	0	-	-
2	WC_FhuA_Deletion Variant	0	-	-
3	WC_FhuA_C1	< 5	n.d.	n.d.
4	WC_FhuA_C2	< 5	n.d.	n.d.
5	WC_FhuA_C3	< 5	n.d.	n.d.
6	WC_FhuA_C1	19	58/42	190
7	WC_FhuA_C2	15	58/42	150
8	WC_FhuA_C3	10	57/43	100

^{a} c(**16**) = 0.025 M. ^{b} Determined by ^{1}H NMR spectroscopy.

The whole-cell itself is not able to catalyze the ROMP of substrate **16** (Table 20, entry 1). In addition, when C545 is absence, the catalyst cannot perform the thiol-maleimide coupling and therefore catalytic activity was not observed (Table 20, entry 2). The latter proves that the washing procedure is sufficient to remove unbound catalysts. If the water concentration was adjusted by vapor diffusion, low conversion was observed and reproducibility of the results is hard to achieve (Table 20, entries 3-5). Applying direct swelling in buffer, the results seem promising. Up to 19% conversion was observed for the shortest linker of catalyst **C1** (Table 20, entry 6). For the longer linker **C2** and **C3** the conversion was 15% and 10%, respectively (Table 20, entries 7 and 8). These results mirror the findings from the soluble protein, where the shortest linker based **FhuA_C1** was the most active.[11] The activity of the whole-cell in general is lower, which can be explained by the structure of the new catalysts. In solution, the monomer can enter from two sides into the channel. In the whole-cell, only one side is accessible. Also, the growing polymer might block the channel at a relatively early stage. Prediction in which direction – in the cell or out of the cell – the polymer will grow, is challenging, since references are not known for this polymer in such an environment.

The obtained *cis*/*trans* ratio of ca. 60/40 is identical to that obtained with the homogenous system, supporting the refolded structure assumed for the soluble system. The tertiary β-barrel structure in the whole-cell is not disturbed in the membrane.

These results show that the **FhuA** based metathease system can be eveolved to whole-cell systems based on *E. coli* in which **FhuA** is located in the outer membrane the natural environment of this protein. The catalytic activity with TON up to **190** seems promising in the ROMP reaction of oxanorbornene **16**. Extending the application toward other methods of small molecule transformation like RCM or CM should make it attractive for organic synthesis or for industrial application.

B.3.3. Cell-Surface Display Technology to Generate a Whole-Cell System Based on Nitrobindin

To construct a whole-cell catalyst based on **NB**, Prof. Schwaneberg and his coworker M.Sc. Alexander R. Grimm developed a display strategy.[13-14] The soluble protein **NB** can be displayed on the cell-surface of *E. coli*. The *E. coli* display system based on an inactive esterase was reported by Kolmar and coworkers.[14] Figure 37 visualizes the **CSD_NB4** with **NB4** as displayed protein.

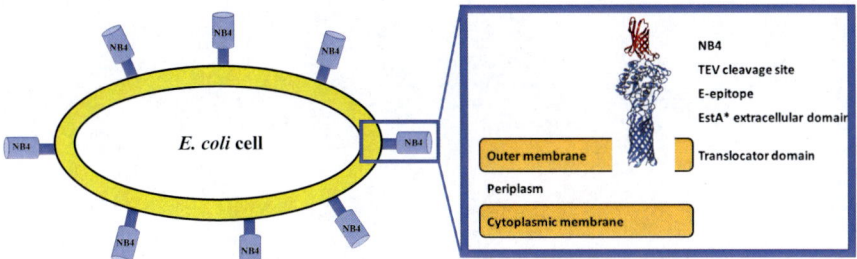

Figure 37. Visualization of the **CSD_NB4**.

To show the feasibility of this strategy for the construction of artificial metalloenzymes, the rhodium-based catalyst of Hayashi and coworkers that polymerizes phenylacetylene was chosen.[9-10] The highly *trans*-selective **NB4** variant was immobilized.[10] Rhodium catalyst **25** was attached to **CSD_NB4** using comparable strategy than for **FhuA_WC** (Scheme 20).

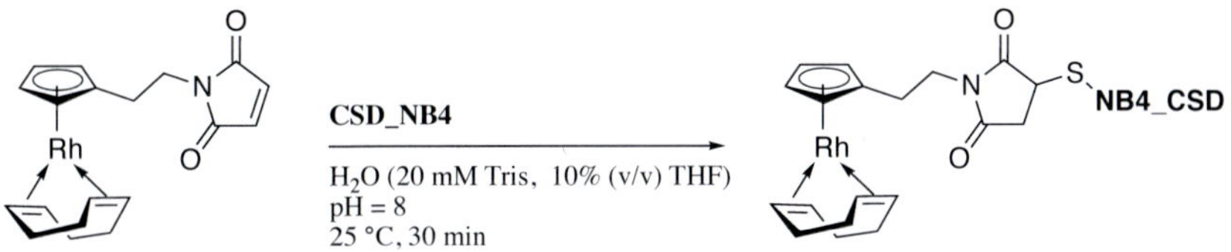

Scheme 20. Conjugation of catalyst **25** to **CSD_NB4**.

Coupling conditions were applied as previously reported by Hayashi and coworkers.[9] The conjugation time was restricted to 30 min, since the protein precipitates from homogeneous systems at longer reaction times.[9] After 30 min, the suspension was centrifuged, the supernatant discarded and excess catalyst was removed by washing with THF. Finally, THF was removed by washing with buffer solution. The generated **CSD_NB4_Rh** was treated with a TEV protease to cleave the **NB4** fragment containing the rhodium catalyst from the surface of *E. coli*. As reference, a sample without the rhodium cofactor was analyzed. The cleaved fragments were analyzed with MALDI-TOF MS. (Figure 38).

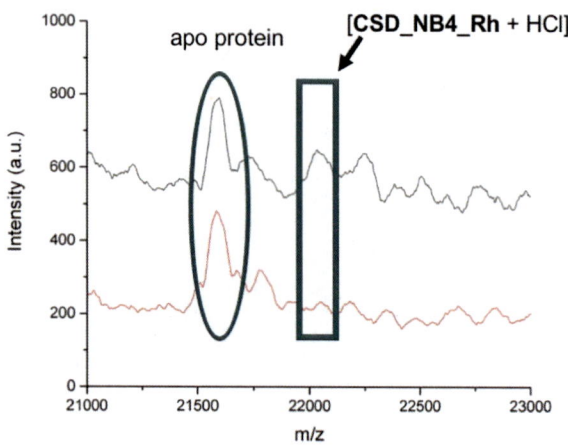

Figure 38. MALDI-TOF MS spectra for **CSD_NB4** (red) and **CSD_NB4_Rh** (black).

The mass for the apoprotein was clearly observed. The **NB4** protein fused to the phoA sequence (responsible for the translocation of the fusion protein through the *E. coli* cytoplasmic membrane) corresponds to a calculated mass of 21581.7 Da. The observed mass with 21582.4 Da is in good agreement with the predicted value (Figure 38, circled signal). For the conjugate **CSD_NB4_Rh**, the mass of apoprotein mass also appeared. The calculated mass for the conjugate is 21981.3 Da. The observed peak at 22017.7 Da corresponds to the HCl adduct of **CSD_NB4_Rh** (Figure 38, black spectrum). This result indicates successful anchoring to the **NB4** protein on the surface of *E. coli*.

The generated biohybrid whole-cell catalyst was tested in the polymerization of phenylacetylene and compared to the results with the homogeneous protein system based on **NB** reported by Hayashi.[9-10] Selected results are shown in Table 21.

Table 21. Selected results for the polymerization of phenylacetylene **33** using **CSD_NB4**.

Entry	Catalyst	Cis/trans[a]	M_n[b] [g/mol]	PDI[b]
1	Tris buffer	-	-	-
2[10]	CSD_NB4	-	-	-
3[10,c]	NB4_Rh	18/82	38900	2.4
4[10]	25	93/7	22900	2.6
5[10]	Mb(A125C)_Rh	91/9	46500	2.4
6	CSD_NB4_Rh	20/80	5500	2.9
7[d]	CSD_NB4	50/50	11000	1.6
8[c]	CSD_NB4	-	-	-
9[e]	pET22	-	-	-
10	pET22 + 25	-	-	-
11	CSD_NB4_C96Q	-	-	-
12	CSD_NB4_C96Q	-	-	-

[a] Determined by ^{1}H NMR spectroscopy in CDCl$_3$ (cf. Figure 39). [b] Determined by GPC. [c] Catalysis in THF. [d] Catalysis in neat **33**. [e] pet22 was used.

Activation of alkynes bearing a Michael acceptor by **FhuA** have been reported previously.[15] To exclude product formation or side reactions with either the buffer medium or the **CSD_NB4**, blank runs were performed (Table 21, entries 1-2). In both reactions, conversion of **33** was not observed. The results were compared to those obtained with the soluble protein **NB4** with Rh cofactor **25** reported by Hayashi and coworkers (Table 21, entries 3-5).[10] Without a defined protein environment, the selectivity of the metal site is the same in water as in THF, favoring the *cis* polymer with more than 90% (Table 21, entries 4-5).[10] The **NB4_Rh** catalyst with a defined cavity favors in contrast the *trans* polymer with 82% (Table 21, entry 3).[10] The whole-cell catalyst **CSD_NB4_Rh** shows the same trend in selectivity, favoring clearly *trans* polymer (80%, Table 21, entry 6). This result indicates the feasibility of the display technology to generate whole-cell catalysts for artificial metalloenzymes. The result shows that these artificial metalloenzymes are suited for whole-cell catalysis. The advantage of whole-cell catalysis will expand the toolbox for artificial biohybrid conjugates.

CSD_NB4_Rh was used in pure phenylacetylene **33**. The color change indicated polymer formation. During the reaction, the initial suspension turned slowly into a viscous, homogeneous mixture. The substrate phenylacetylene seems to dissolve the cells and denature the protein structure. This explains why *trans*-selectivity was not observed (cis/trans of 50/50, Table 21, entry 7). Nevertheless, the reaction proceeds. Further stabilization of the cell e.g. by incorporation into a microgel should support the whole-cell conjugate. In pure THF, the suspension remained heterogeneous, but no polymer formation was observed (Table 21, entry 8). This is explained by a loss of the protein structure of **NB4** on the cell-surface what leads to a deactivation of the rhodium site. Further blank runs were performed with an empty vector cell that did not contain the target protein **NB4,** either with or without catalyst **25.** No polymer was formed (Table 21, entries 9-10). The latter proves that the washing procedure is sufficient to remove unbound catalysts. This was also observed, if the cysteine at position C96 that binds the catalyst had been mutated back to the native glutamine C96Q (Table 21, entries 11-12). A mixture of one equivalent of **25** with the **CSD_NB4** did not give any conversion, since the catalyst seems to be bound in an unspecific manner and is deactivated.

B.3.4. Summary and Conclusion

The transmembrane protein **FhuA** allows to incorporate catalytic centers within the outer membrane of *E. coli* to serve as an artificial metathease. The whole-cell conjugates **WC_FhuA_C1-3** catalyzed the ROMP reaction of substrate **16**. Even though the conversion is not as high as with the analogous systems based on the solubilized protein, the selectivity is the same. These results prove that whole-cell catalysis is possible by utilizing a membrane protein Therefore, this approach fundamentally different to the whole-cell system presented by Ward and coworkers.[8]

The artificial metalloprotein **NB4_Rh** was immobilized on a cell-surface and catalyzed the polymerization of phenylacetylene. This strategy made artificial whole-cell based catalysts available for proteins that do not naturally occur on a cell-surface or within the cell membrane. The results show clearly that the defined protein environment including the metal cofactor can be transferred to the surface of the cell and be used as whole-cell catalyst.

B.3.5. Experimental Section

The proteins used in this chapter were designed and provided by the group of Prof. Schwaneberg at RWTH Aachen University and his coworker M.Sc. Alexander Grimm.

All operations were performed under an inert atmosphere of argon or nitrogen using standard Schlenk or glove box techniques. Water and other solvents were degassed by using "freeze-pump-thaw" technique. Dichloromethane and tetrahydrofuran were obtained dry and degassed from a SPS 800 from *MBraun*. Chloroform-d_1 was dried over calcium hydride, distilled, degassed and stored in a glove box. NMR-Spectra were recorded on a Bruker DRX 400 spectrometer (^{1}H, 400.1 MHz). Chemical shifts were referenced internally by using the residual solvent resonances.[16] MALDI–TOF MS spectra were recorded on an Ultraflex III TOF/ TOF mass spectrometer (*Bruker Daltonics*). Compounds **C1**,[11] **C2**,[11] **C3**,[1] 7-Oxanorbornene derivative **16**[17-18] and rhodium complex **25**[9] were synthesized according to literature procedures. Phenylacetylene **33** is commercially available and was used as received. Digestion with a TEV protease was performed by the group of Prof. U. Schwaneberg at the Institute for Biotechnology at RWTH Aachen University. All other chemicals were used as received if not mentioned otherwise.

General procedure for the conjugation to WC_FhuA.

The lyophilized **WC_FhuA** (ca. 250 mg) was swollen in an aqueous solution (10 ml, 100 mM NaCl, pH = 8, adjusted with NaOH) for 48-68 h. The mixture was centrifuged, the supernatant discarded and the pallet resuspended in water (5 ml, pH = 8 adjusted with NaOH). Catalyst **C1-C3** (10 equiv.) was added dropwise in THF (10% (v/v)). The solution was incubated for 16 h at 25 °C. The mixture was centrifuged, the supernatant discarded and the pallet was washed with an aqueous solution containing NaP$_i$ (10 mM, pH = 8.0) and THF (50% (v/v)). The mixture was centrifuged and the supernatant discarded. This step was repeated twice, in total three times washing. Afterwards, the cell pallet was resuspended in the catalysis buffer consisting of an aqueous NaP$_i$ buffer (4 mL, 100 mM, pH = 5.8, 50 mM NaCl).

General procedure for the catalysis using WC_FhuA.

To the freshly prepared **WC_FhuA_C1-3**, THF (10% (v/v)) was added and incubated for 15 minutes at 25 °C. Substrate **16** (15 µL) was added via micro syringe. The mixture was slowly stirred for 24 h at 25 °C. The reaction was quenched with ethyl vinyl ether (100 equiv.) in THF (5 mL). The mixture was dried over MgSO$_4$ and filtered. The solvent was *removed in vacuo*, the residue taken up in CDCl$_3$ and analyzed by ^{1}H NMR spectroscopy as described by Feast and Harrison.

General procedure for the conjugation to CSD_NB4.

The lyophilized CSD_NB4 (ca. 250 mg) was swollen in an aqueous solution (10 ml, 50 mM NaCl, 10 mM Tris buffer, pH = 8.0) for 48-68 h. The mixture was centrifuged, the supernatant discarded and the pallet resuspended in aqueous buffer solution (10 ml, 20 mM Tris buffer, pH = 8.0). Catalyst **25** (10 equiv.) was added dropwise in THF (10% (v/v)). The solution was incubated for 30 min at 25 °C. The mixture was centrifuged, the supernatant discarded and the pallet was washed with an aqueous solution containing Tris (10 mM, pH = 8.0) and THF (50% (v/v)). The mixture was centrifuged and the supernatant discarded. This step was repeated twice, in total three times washing. Afterwards, the cell pallet was resuspended in the catalysis buffer consisting of an aqueous Tris buffer (2 mL, 10 mM, pH = 8.0).

General procedure for the catalysis using WC_FhuA.

To the freshly prepared **CSD_NB4_Rh**, substrate **16** (330 µL) was added via syringe. The mixture was slowly stirred for 12 h at 25 °C. The solvent was removed *in vacuo*, the polymer extracted with THF, dried over MgSO4 and filtered. The solvent was *removed in vacuo*, the residue taken up in CDCl$_3$ and analyzed by ^{1}H NMR spectroscopy.

Analysis of the polymer produced with CSD_NB4_Rh

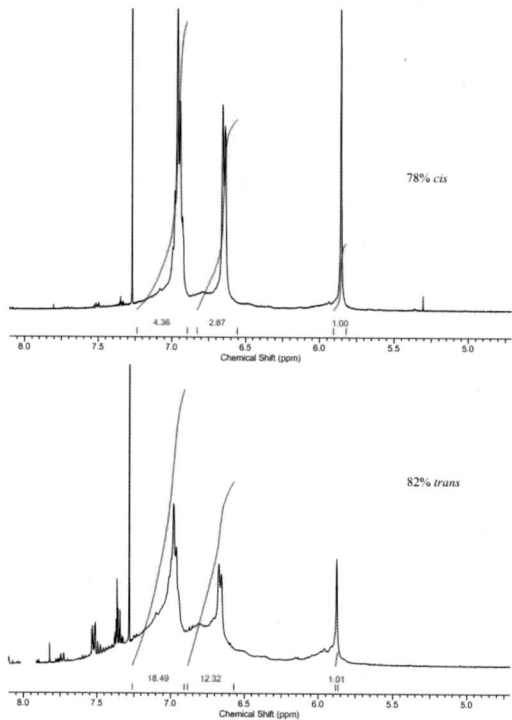

Figure 39. ^{1}H NMR spectroscopy (400 MHz, CDCl$_3$, 25 °C) of poly(phenylacetylene) **34**.

B.3.6. References

(1) Philippart, F.; Arlt, M.; Gotzen, S.; Tenne, S.-J.; Bocola, M.; Chen, H.-H.; Zhu, L.; Schwaneberg, U.; Okuda, J. *Chem. Eur. J.* **2013**, *19*, 13865.

(2) Bianchetti, C. M.; Blouin, G. C.; Bitto, E.; Olson, J. S.; Phillips, G. N. *Proteins: Struct., Funct., Bioinf.* **2010**, *78*, 917.

(3) Ersson, B.; Rydén, L.; Janson, J.-C. In Protein Purification; John Wiley & Sons, Inc.: **2011**, p 1.

(4) Schmid, A.; Dordick, J. S.; Hauer, B.; Kiener, A.; Wubbolts, M.; Witholt, B. *Nature* **2001**, *409*, 258.

(5) Menzel, A.; Werner, H.; Altenbuchner, J.; Gröger, H. *Eng. Life Sci.* **2004**, *4*, 573.

(6) Jakoblinnert, A.; Mladenov, R.; Paul, A.; Sibilla, F.; Schwaneberg, U.; Ansorge-Schumacher, M. B.; de Maria, P. D. *Chem. Commun.* **2011**, *47*, 12230.

(7) Thomas, S. M.; DiCosimo, R.; Nagarajan, V. *Trends Biotechnol.* **2002**, *20*, 238.

(8) Jeschek, M.; Reuter, R.; Heinisch, T.; Trindler, C.; Klehr, J.; Panke, S.; Ward, T. R. *Nature* **2016**, *537*, 661.

(9) Onoda, A.; Fukumoto, K.; Arlt, M.; Bocola, M.; Schwaneberg, U.; Hayashi, T. *Chem. Commun.* **2012**, *48*, 9756.

(10) Fukumoto, K.; Onoda, A.; Mizohata, E.; Bocola, M.; Inoue, T.; Schwaneberg, U.; Hayashi, T. *ChemCatChem* **2014**, *6*, 1229.

(11) Sauer, D. F.; Bocola, M.; Broglia, C.; Arlt, M.; Zhu, L.-L.; Brocker, M.; Schwaneberg, U.; Okuda, J. *Chem. Asian J.* **2015**, *10*, 177.

(12) Zhu, L.; Arlt, M.; Liu, H.; Bocola, M.; Sauer, D. F.; Gotzen, S.; Okuda, J.; Schwaneberg, U. In *Bio-Synthetic Hybrid Materials and Bionanoparticles: A Biological Chemical Approach Towards Material Science*; The Royal Society of Chemistry: **2015**, p 57.

(13) Yang, T. H.; Pan, J. G.; Seo, Y. S.; Rhee, J. S. *Appl. Environ. Microbiol.* **2004**, *70*, 6968.

(14) Becker, S.; Theile, S.; Heppeler, N.; Michalczyk, A.; Wentzel, A.; Wilhelm, S.; Jaeger, K.-E.; Kolmar, H. *FEBS Lett.* **2005**, *579*, 1177.

(15) Shen, R.; Sauer, D. F.; Okuda, J. *unpublished results* **2017**.

(16) Fulmer, G. R.; Miller, A. J. M.; Sherden, N. H.; Gottlieb, H. E.; Nudelman, A.; Stoltz, B. M.; Bercaw, J. E.; Goldberg, K. I. *Organometallics* **2010**, *29*, 2176.

(17) Novak, B. M.; Grubbs, R. H. *J. Am. Chem. Soc.* **1988**, *110*, 960.

(18) Feast, W. J.; Harrison, D. B. *Polym. Bull.* **1991**, *25*, 343.

C. Summary

The work presented here describes the design, synthesis, characterization and application of biohybrid conjugates consisting of a Grubbs-Hoveyda type unit in **FhuA** or nitrobindin as protein scaffolds. Furthermore, the successfully transition to whole-cell system for both proteins was shown.

In chapter B1, the metathease based on the transmembrane protein FhuA was investigated in more detail. New Grubbs-Hoveyda (GH-type) catalysts **C1-C3** varying in the linker length were synthesized along with the corresponding cysteine conjugates. These catalysts and cysteine conjugates were analyzed by ESI-TOF MS and the results compared to the biohybrid conjugates **FhuA_C1-3**. In the ROMP reaction of 7-oxanorbornene derivative 3,4-Bis(methoxymethyl)-7-oxanorbornene (**16**), the attached cysteine does not cause any change in selectivity (*cis/trans* 50/50 for catalyst precursor **COH** and **C1-3_Cys**). Likewise, the biohybrid conjugates based on **FhuA** did not show any influence (58/42 for **FhuA_C1-C3**).

Three different solubilizing reagents were investigated in this study. The refolding reagents 2-Methyl-2,4-pentanediol (MPD) and polyethylene-poly(ethylene glycol) (PE-PEG) and the surfactant sodium dodecyl sulfate (SDS) were applied with the biohybrid catalysts **FhuA_C1-3**. In the ROMP reaction of 7-oxanorbornene **16**, a change in activity and selectivity was not observed between the refolding reagents MPD and PE-PEG. The small molecular refolding reagent MPD can be utilized for substrates that would interact strongly with the micelles formed by PE-PEG. This feature makes the membrane protein based system more flexible in terms of substrate choice. SDS disturbs the β-barrel structure what accelerates the ROMP and RCM reaction.

The activity of the FhuA-based metathease in cross metathesis was investigated, revealing high chemoselective cross-metathesis of 1-buten-4-ol, without formation of isomerization products.

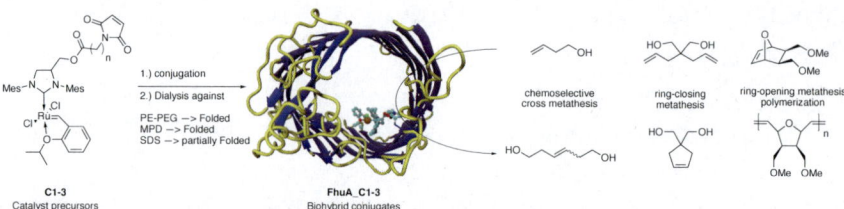

Scheme 21. Summary of Chapter B1. Catalyst precursors **C1-3** were synthesized and covalently attached to the protein **FhuA**. The biohybrid conjugates **FhuA_C1-3** were tested in either PE-PEG, MPD or SDS in different metathesis reactions.

In chapter B2, the use of the small β-barrel protein nitrobindin as scaffold for artificial metatheases was investigated. Initially, the study was conducted with two mutants varying in cavity size. The **NB4** mutant, having a smaller cavity of 855 Å^3, could only achieve poor coupling efficiencies with the longest spacer of catalyst **C3**. The **NB1** mutant offers ~30% more cavity volume compared to **NB4** and performs coupling excellently with the **C3** catalyst, moderately with **C2**, and poorly with **C1**. All three catalysts could be attached within the nitrobindin scaffolds. The cavity volume seems to be the crucial factor for proper incorporation. To generate more space within the cavity, a new approach was introduced in cooperation with the group of Prof. Schwaneberg (performed by M.Sc. Alexander R. Grimm). The barrel extension of nitrobindin allowed to mutate the **NB4** mutant to the **NB4exp** mutant by duplication of two β-strands, leading to almost doubled cavity volume. The bulky catalysts **C1**-**C3** could all be incorporated with excellent coupling efficiencies. The advantage of the larger cavity can be seen in the activity of the conjugates bearing the shorter spacer. They show a relatively high activity compared to the tighter cavity variant **NB4**. In contrast to the dimeric structure for **NB4** or **NB1**, **NB4exp** is monomeric. This allows the use of hydrophobic substrates in the RCM and CM reaction. Additionally, the larger distance of the catalyst to the rim of the cavity moves the positively charged lysines away from the active site, allowing the use of positively charged substrates. High-throughput site-directed mutagenesis becomes possbile by a fluorescence screening assay.

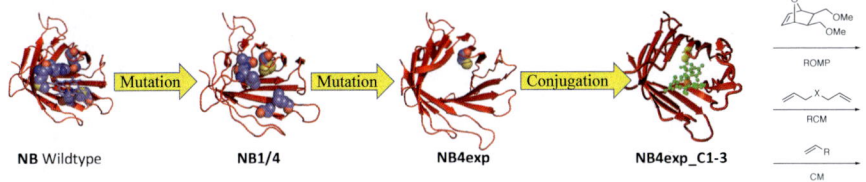

Scheme 22. Development of an artificial metathease based on **NB**.

In chapter B3, two new strategies for the construction of artificial whole-cell catalysts were described. The natural location of the **FhuA** in the outer membrane of *E. coli* was used to construct an artificial metathease as whole-cell catalyst. The activity in ROMP was slightly reduced compared to that of the homogeneous protein, while the selectivity retained, showing the feasibility of this approach. This indicates that the homogeneous protein has the same tertiary structure compared to the whole-cell FhuA **WC_FhuA** system. The construction of an artificial whole-cell system based on nitrobindin was realized by using the surface displayed variant of **NB4** on the surface of *E. coli*. The generated **CSD_NB4_Rh** catalyst was evaluated in the polymerization of phenylacetylene, reported previously by Hayashi and coworkers for the homogeneous **NB4_Rh** system. The selectivity within the polymer chain was retained and the heterogeneous system behaves like the hybrid catalysts in solution. This indicates the usefulness of this approach to generate whole-cell catalysts and expands the toolbox for the construction and design of artificial metalloenzymes.

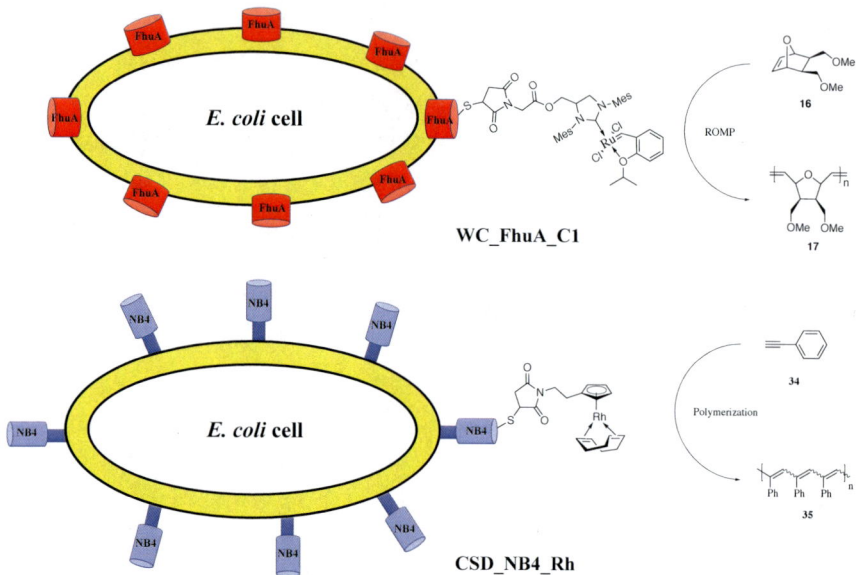

Scheme 23. Whole-cell systems synthesized in this work. Top: Whole-cell catalysts based on **FhuA** located in the outer membrane of *E. coli*. The GH-type catalyst was attached and the catalyst **WC_FhuA_C1** was tested in the ROMP reaction of substrate **16**. Bottom: Whole-cell catalyst based on nitrobindin, attached to the surface of *E. coli via* a cell surface display strategy. [Rh] catalyst **25** was attached, and the catalyst **CSD_NB4_Rh** was used in the polymerization of phenylacetylene **34**.

D. Appendix

D.1. List of Abbreviation

Avi	avidin
CD	circular dichroism
CM	cross metathesis
DMSO	dimethyl sulfoxide
ESI-TOF	electrospray ionization – time of flight
FhuA	*Ferric hydroxamate uptake protein component: A* used as abbreviation for the **FhuA_ΔCVF^{tev}** mutant
hCAII	human carbonic anhydrase type II
ICP-AES	inductively coupled plasma – atomic emission spectroscopy
MALDI-TOF	matrix-assisted laser desorption/ionization – time of flight
MjHSP	*M. jannashii small heat shock protein 16.5*
MS	mass spectrometry
NB	nitrobindin
NHC	*N*-heterocyclic carbene
NMM	*N*-methyl morpholine
NMR	nuclear magnetic resonance
PDB	protein data base
P_i	inorganic phosphate
RCM	ring-closing metathesis
ROMP	ring-opening metathesis polymerization
Sav	streptavidin
SDS	sodium dodecyl sulfate
TEV	tobacco etch virus
TON	turnover number
WT	wild-type
δ	chemical shift in ppm

D.2. Curriculum Vitae

Personal Data

Daniel Friedrich Sauer

Address:	Reumontstraße 36A, 52064 Aachen, Germany
Date and Place of Birth:	06.01.1988; Düren, Germany
Email address:	daniel.sauer@rwth-aachen.de

Education

Graduate Studies

10.2013 – present   Doctoral studies at RWTH Aachen, Germany, under supervision of Prof. Jun Okuda (SeleCa Fellowship)

Undergraduate Studies

4. - 9.2013   Master thesis in the group of Prof. Dr. Okuda, Institute of Inorganic Chemistry, RWTH Aachen University Title: *"Biohybrid Catalysts for Olefin Metathesis"*

Master examination with the overall mark *"excellent"*

5. - 8. 2011   Bachelor thesis in the group of Prof. Dr. Rueping, Institute of Organic Chemistry, RWTH Aachen University Title: *"Synthesis and Application of Photoredoxcatalysts Constructed with Quinoline Ligands"*

10.2007 – 9.2013   Chemistry studies at RWTH Aachen University, Germany; Graduation B. Sc. RWTH and M. Sc. RWTH

Secondary Education

06.2007   Qualification for university entrance (Abitur),

Gymnasium am Wirteltor, Düren.

D.3. Publications

D.3.1. Publications in Peer-Reviewed Journals

(1) Sauer, D. F.; Bocola, M.; Broglia, C.; Arlt, M.; Zhu, L.; Brocker, M.; Schwaneberg, U.; Okuda, J.,
Chem. Asian J. **2015**, *10*, 177:
Hybrid Ruthenium ROMP Catalysts Based on an Engineered Variant of β-Barrel Protein FhuA ΔCVFtev: Effect of Spacer Length

(2) Sauer, D. F.; Himiyama, T.; Tachikawa, K.; Fukumoto, K.; Onoda, A.; Mizohata, E.; Inoue, T.; Bocola, M.; Schwaneberg, U.; Hayashi, T.; Okuda, J.
ACS Catal. **2015**, *5*, 7519:
A Highly Active Biohybrid Catalyst for Olefin Metathesis in Water: Impact of a Hydrophobic Cavity in a β-Barrel Protein

(3) Himiyama, T.; Sauer, D. F.; Onoda, A.; Spaniol, T. P.; Okuda, J.; Hayashi, T.
J. Inorg. Biochem. **2016**, *158*, 55:
Construction of a hybrid biocatalyst containing a covalently-linked terpyridine metal complex within a cavity of aponitrobindin

(4) Sauer, D. F.; Gotzen, S.; Okuda, J.
Org. Biomol. Chem. **2016**, *14*, 9174:
Metatheases: artificial metalloproteins for olefin metathesis

(5) Osseili, H.; Sauer, D. F.; Beckerle, K.; Arlt, M.; Himiyama, T.; Polen, T.; Onoda, A.; Schwaneberg, U.; Hayashi, T.; Okuda, J.
Beilstein J. Org. Chem. **2016**, *12*, 1314:
Artificial Diels–Alderase based on the transmembrane protein FhuA

(6) Mukherjee, D.; Sauer, D. F.; Zanardi, A.; Okuda, J.
Chem. Eur. J. **2016**, *22*, 7730:
Selective Metal-Free Hydrosilylation of CO_2 Catalyzed by Triphenylborane in Highly Polar, Aprotic Solvents

D.3.2. Book Contributions

(1) Zhu, L.; Arlt, M.; Liu, H.; Bocola, M.; Sauer, D. F.; Gotzen, S.; Okuda, J.; Schwaneberg, U. In Bio-Synthetic Hybrid Materials and Bionanoparticles: A Biological Chemical Approach Towards Material Science; The Royal Society of Chemistry: 2015, p 57.

D.3.3. Patent Application

(1) Schwaneberg, U., Okuda, J., Sauer, D.F., Arlt, M., Zhu, L.; Bocola, M., Grimm, A.R.; Mertens, A., Ganzzell-basierte Biohybridkatalysatoren für selektive chemische Umsetzungen; Deutsche Patentanmeldung, 22.12.2016, 102016125516.6

D.3.4. Conference Contributions

(1) "FhuA Based Grubbs-Hoveyda Type Catalysts for Ring Opening Metathesis Polymerization", Biotechnology and Chemistry for Green Growth, March 2014, Osaka, Japan, oral presentation.

(2) "Nitrobindin Based Biohybrid Catalysts for Olefin Metathesis", Biotechnology and Chemistry for Green Growth, March 2015, Osaka, Japan, oral presentation.

(3) "Nitrobindin Based Artificial Metathease", Biotechnology and Chemistry for Green Growth, March 2016, Osaka, Japan, oral presentation.

(4) "Biohybrid Catalysts for Olefin Metathesis", Aachen-Osaka Joint Symposium within the IRTG "SeleCa", September 2014, Aachen, Germany, oral presentation.

(5) "β-Barrel Protein as Scaffold for Biohybrid Catalysis: Activation of C-C Triple Bonds", Aachen-Osaka Joint Symposium within the IRTG "SeleCa", September 2015, Aachen, Germany, oral presentation.

(6) "Construction of Artificial Metatheases", Workshop on Artificial Metalloenzymes, September 2016, Aachen, Germany, oral presentation.

(7) "Biohybrid Catalysts Incorporated in a β-Barrel Protein for Olefin Metathesis in Water", The 96th Annual Meeting of the Chemical Society of Japan, March 2016, Kyoto, Japan, oral presentation.

(8) "Biohybrid Grubbs-Hoveyda Type Catalysts for Ring Opening Metathesis Polymerization", 26th International Conference on Organometallic Chemistry (July 2014), Sapporo, Japan, poster presentation.

(9) "β-Barrel Protein as Scaffold for Biohybrid Catalysis", 13th International Symposium on Applied Bioinorganic Chemistry (June 2015), NUI Galway, Ireland, poster presentation (poster award).

D.3.5. Other Publications

(1) Sauer, D.F.; Schindler, T.
Nachr. Chem. **2017**, *65*, 73:
Deutsch-Japanischer Katalyse-Workshop

D.4. Acknowledgements – Danksagung

Als erstes möchte ich mich bei meinem Doktorvater Prof. Jun Okuda für viele fruchtbare Diskussionen, die Bereitstellung hervorragender Arbeitsbedingungen und die Möglichkeit zur Bearbeitung dieses innovativen und herausfordernden Themas und nicht zuletzt für die Korrektur dieser Arbeit bedanken.

Ebenfalls möchte ich mich bei unseren Koopertionspartnern in Aachen, der Biotechnologiegruppe von Prof. Ulrich Schwaneberg für die vielen Diskussionen und der Bereitstellung der Proteine bedanken. Bei Herrn Prof. Ulrich Schwaneberg bedanke ich mich darüber hinaus für die Übernahme des Zweitgutachtens und für viele Vorschläge zur Verbesserung dieser Arbeit.

I want to thank the group of Prof. Takashi Hayashi at the Osaka University for a memorable time in the laboratory, beside the laboratory and the generous support during my stay in Japan. Especially, I want to thank Prof. Takashi Hayashi, Assoc. Prof. Akira Onoda, Ass. Prof. Koji Oohora for very fruitful discussions and amazing support in Japan. Furthermore, I want to thank Prof. Takashi Hayashi for joining my defense committee. I want to thank Ms. Kijomi Lee for great support during the organization of my stays and for the support in Osaka. I want to thank especially Dr. Tomoki Himiyama, Dr. Yoshitsugu Morita and Mr. Shunsuke Kato for their support in the lab and helping me to get started quickly.

Ich möchte mich ganz Herzlich bei Prof. Sonja Herres-Pawlis für die Übernahme der Funktion als Drittprüferin und bei Prof. Markus Albrecht für die Übernahme der Funktion des Prüfungsvorsitzenden Bedanken.

Ich bedanke mich bei meinen vielen Kollegen der interdisziplinären Biohybridgruppe für viele Diskussionsrunden, großartige Unterstützung, eine tolle Arbeitsatmosphäre und durchweg freundliche Zusammenarbeit.

Ich danke den aktuellen und ehemaligen Mitgliedern des AK Okuda für viele Diskussionen und Hilfestellungen. Ein ganz besonderer Dank gilt Herrn Dr. Klaus „Analytix" Beckerle für die unzähligen Analysen und Hilfestellungen bei analytischen Problemen (ohne die diese Arbeit sicher nicht zustande gekommen wäre), viele fruchtbare Diskussionen und die großartige

Unterstützung, wenn das Projekt etwas ins Stocken geraten ist. Herrn Toni Gossen, Frau Rachida Bohmarat und Herrn Dr. Gerhard Fink danke ich für die Unterstützung bei den NMR Messungen. Herrn Dr. Thomas Spaniol danke ich für die Korrektur dieser Arbeit. Frau Simone Becher danke ich herzlichst für die organisatorischen Hilfestellungen und die Unterstützung beim Kampf mit Bürokratie.

Ich danke meinen Forschungs- und Masterstudenten sowie meinen „HiWis" für die großartige Arbeit, die ihr im Rahmen dieser Arbeit geleistet habt. Ohne euch und eure Beiträge wäre diese Arbeit sicherlich weniger umfangreich geworden.

Mein größter Dank gilt meiner Familie und ganz besonders meinen Eltern. Ihr habt mich sowohl vor, als auch während des Studiums und der Doktorarbeit immer unterstützt und immer an mich geglaubt. Danke!